Fundamentals of Turbulence and Its Modeling

Sawan S. Sinha

Fundamentals of Turbulence and Its Modeling

Ane Books Pvt. Ltd.

Springer

Sawan S. Sinha
Department of Applied Mechanics
Indian Institute of Technology Delhi
New Delhi, Delhi, India

ISBN 978-3-031-94015-6 ISBN 978-3-031-94016-3 (eBook)
https://doi.org/10.1007/978-3-031-94016-3

Jointly published with Ane Books Pvt. Ltd.
In addition to this printed edition, there is a local printed edition of this work available via Ane Books in South Asia (India, Pakistan, Sri Lanka, Bangladesh, Nepal and Bhutan) and Africa (all countries in the African subcontinent).
ISBN of the Co-Publisher's edition: 978-81-97240-36-2

This Springer imprint is published by the registered company Springer Nature Switzerland AG
The registered company address is: Gewerbestrasse 11, 6330 Cham, Switzerland

In the loving memory of my father
Shri Amarnath Suman

Preface

Many flows of engineering interest are turbulent in nature, and thus it is important that students of Aerospace, Mechanical, Chemical, Civil and other allied branches of engineering must be given an opportunity to have at least one-semester long course on the fundamentals of turbulence and its modeling. Some educational institutes in India do offer such a course on turbulent flows and the enrolled students often look for an easy-to understand book on this otherwise difficult subject of study.

Indeed there are some excellent books, written by some renowned authors, already available on this topic. The book *First Course in Turbulence* by H. Tennekes and J. L. Lumley is legendary: it has trained a generation of engineers and researchers worldwide on the subject. In recent years, the book *Turbulent Flows* by S. B. Pope has gained immense popularity owing to the mathematical rigor and the wide range of topics that it covers. With its focus especially on turbulence closure models and their implementations, the book *Turbulence Modeling for CFD* by D. C. Wilcox has been an equally popular book for reference. While these books are indeed extensive in terms of their contents and they do include both fundamental and advanced topics of the subject, it may still be quite challenging for teachers and students to cover most of the contents of any of these books during a single semester-long course. While we all do try to pick-and-choose topics based on their significance, it is often a struggling process for the teacher and the students alike to ascertain what all to cover and with what degree to detail within the time constraints of a semester.

The purpose of this new book is to cover the most essential aspects of turbulence and its modeling such that the included contents can be taught at a pace and in a sequence that it becomes easier for students to learn the matter over a semester-long duration. The author believes that the sequence of contents, the volume of contents and the style of presentation included in this book are unique and apt so that students can make a very smooth transition from their already completed essential training in basic fluid mechanics to their next pursuit of learning advanced, complex but extremely useful concepts related to turbulence and its modeling. This book can ideally be used as the text book to teach turbulence and its modeling over a semester of 12–14 weeks to the final year under-graduate/post-graduate students of Aerospace, Mechanical, Chemical, Civil and other allied branches of engineering. This book, by no means, is being presented or claimed as a substitute to any of the excellent existing books on the topic. However, this book can prove to be a useful stepping stone preparing a student to reach a stage where reading and comprehending other advanced texts and research papers related to this field of study would become easier.

This book is based on my lecture notes that I have used to teach the course *Turbulence and Its Modeling* several times at Indian Institute of Technology Delhi. In fact, my decision to invest effort and time to convert those notes into a book is based on the persistent encouraging feedback from my students about the choice of contents and the organization of the course that I have offered in the past semesters.

I would like to thank my current and former students Shishir Srivastava, Deep Shikha, Sagar Saroha and Farooq Ahmad Bhat for their generous efforts in helping me gather data, prepare figures and for providing constructive feedback at various stages of writing this book. Special thanks are due to Shishir for creating the cover picture for this book.

New Delhi, India Sawan S. Sinha

Competing Interests The author has no competing interests to declare that are relevant to the content of this manuscript.

Contents

1 **Tensors** 1
1.1 Expressing a First-Order Tensor Using a Cartesian Coordinate System 2
1.2 Transformation Rule: First-Order Tensors 2
1.3 Expressing a Second-Order Tensor Using a Cartesian Coordinate System 4
1.4 Transformation Rule: Second-Order Tensors 6
1.5 Expressing Higher-Order Tensors Using a Cartesian Coordinate System 6
1.6 Einstein's Summation Rule 7
1.7 Tensor Operations 9
1.7.1 Dot Product of Two Tensors 9
1.7.2 Double Dot Product of Two Tensors 12
1.7.3 Cross Product of Two Vectors 12
1.8 The ϵ-δ Identity 13
1.9 Spatial Derivatives of Tensors 13
1.10 Index Notation and Tensor Identities 15

2 **Description of Fluid Kinematics** 19
2.1 The Continuum Description 19
2.2 The Lagrangian Description of Fluid Continuum 20
2.3 The Eulerian Description of Fluid Continuum 22
2.4 Kinematics of a Fluid Element in a 3C, 3D Velocity Field 24
2.5 A Vortex in a Flow Field 27
2.6 Characteristic Length and Time-Scales Associated with an Eddy 28
2.7 An Idealized Superposition of Eddies 29

3 **Nature of Turbulent Flows** 33
3.1 Governing Equations of Turbulent Flows 36

4 **Random Variables and Their Characterization** 39
4.1 An Event 39
4.2 Probability of an Event 40

4.3 Cumulative Distribution Function of a Random Variable 41
4.4 Probability Density Function of a Random Variable 41
4.5 Some Examples of Known PDFs 42
4.6 Joint CDF of Multiple Random Variables 43
4.7 Joint PDF of Multiple Random Variables 44
4.8 Expected Value of a Function of Random Variables 45
4.8.1 Expectation of the Product of Q and a Non-random Quantity a 45
4.8.2 Expectation of the Expectation of Q 46
4.8.3 Expected Value of a Random Variable ϕ_1 46
4.8.4 Expected Value of a Sum of Random Variables 47
4.8.5 Fluctuation in a Random Variable ϕ_1 48
4.8.6 Expectation of the Product of Two Random Variables 49
4.8.7 Moments of a Random Variable 49
4.9 Expectation of Derivatives of Random Processes and Random Fields 51
4.9.1 Partial Derivative with Respect to Time 51
4.9.2 Partial Derivative with Respect to Spatial Coordinates 52
4.10 Categorization of Turbulent Flows Based on Their Statistics 52
4.11 Expectation and Averaging 53
4.11.1 Ensemble Averaging 53
4.11.2 Time-Averaging 53
4.11.3 Line Averaging 54
4.11.4 Area Averaging 54
4.11.5 Volume Averaging 55

5 Governing Equations of the Mean Flow Field 57
5.1 The Mean Continuity Equation 57
5.2 The Mean Momentum Equation 58
5.3 The Turbulence Closure Problem 59
5.4 The Reynolds Stress Tensor 60
5.5 Transport Equation of the Reynolds Stress Tensor 62
5.6 Turbulence Kinetic Energy 69
5.7 The RSTE and the Turbulence Closure Problem 74

6 Turbulence Near a Solid Wall 77
6.1 Observed Behavior of the Mean Flow Field 77
6.2 Simplified Governing Equations of the Mean Flow Field 80
6.3 Order-of-Magnitude Analysis of the Governing Equations 82
6.3.1 Continuity Equation 86
6.3.2 The $\langle V_1 \rangle$ Equation 87
6.3.3 The $\langle V_2 \rangle$ Equation 90

6.4 Anatomy of a Flat Plate Turbulent Boundary Layer 96
6.5 Near-Wall Asymptotic Behavior of the Fluctuating Velocity Vector 101

7 Understanding Multiplicity of Length-Scales in Turbulent Flows 105
7.1 Fourier Description of a Spatially Periodic Function 107
7.2 Spectral Density Functions of Turbulence Kinetic Energy and Its Dissipation Rate 116
7.3 Evolution of Energy and Dissipation Spectra in Decaying Turbulence: DNS-Based Observations 120
7.4 Explanation of Energy Cascade: Fourier Description of Navier-Stokes Equation 123
7.5 Kolmogorov's Hypotheses 127
7.5.1 Kolmogorov's First Similarity Hypothesis 130
7.5.2 Kolmogorov's Second Similarity Hypothesis 131
7.5.3 Kolmogorov's Hypothesis of Local Isotropy 133

8 Turbulence Modeling 135
8.1 Eddy Viscosity Closure 136
8.2 Modeling the Unclosed Terms: A Broad Outline 140
8.3 The k-ω Model: A Case Study on Turbulence Modeling 141
8.3.1 Selection of Closure Coefficients 146
8.4 The k-ϵ Model: A Case Study on Turbulence Modeling 153
8.4.1 Selection of Closure Coefficients 156

9 Scale-Resolving Simulations of Turbulent Flows 165
9.1 The Filtered Description of a Turbulent Flow Field 166
9.2 Governing Equations of a Filtered Flow Field 167

Bibliography 171

Index 173

About the Author

Dr. Sawan S. Sinha is a Professor in the Department of Applied Mechanics, Indian Institute of Technology (IIT) Delhi, India. He received his B.Tech. degree in Aerospace Engineering from IIT Kanpur (2001) and his Ph.D. degree from Texas A&M University (Aerospace Engineering), College Station, TX, USA (2009). Besides being a faculty member at IIT Delhi, his professional experience from the past includes working as an engineer for Defence Research Development Organization (DRDO) during 2003–2004 and later as a post-doctoral Research Associate at Texas A&M University (2010–2012).

Dr. Sinha has more than a decade-long teaching experience at IIT Delhi. Besides teaching *Turbulence and Its Modeling*, he frequently teaches courses on Fluid Mechanics, Engineering Mechanics, and Computational Fluid Dynamics, as well. Based on students' feedback, he has received Awards for Teaching Excellence at IIT Delhi (Turbulent Shear Flows, 2013 and Fluid Mechanics, 2014).

1 Tensors

A *tensor of order n* ($\in \{0, 1, 2, 3, ...\}$) in three-dimensional space is defined as a mathematical entity consisting of 3^n components. The numerical values of these components can vary depending on the coordinate system used to represent the tensor. These 3^n numbers are called the *scalar components of the tensor* within the chosen coordinate system. Although these scalar components generally depend on the chosen coordinate system for their representation, the tensor itself—as a mathematical entity and as a representation of a physical quantity—remains independent of the coordinate system used. In other words, the tensor itself is *invariant* to the choice of the coordinate system used to express it. This property of invariance is ensured by a set of relationships between the 3^n components of the tensor in one coordinate system and the corresponding set of 3^n components in another coordinate system. These relationships ensure that despite the different scalar representations, the tensor remains consistent across all coordinate systems. Such a relationship is called the *transformation rule* of all tensors of order n. A tensor of order zero is a special case wherein the transformation rule is trivial. Such a tensor is described by one (3^0) number, which is independent of the choice of the working coordinate system.

In our study of fluid mechanics and turbulence, we come across tensors of various orders of fluid. Some examples of tensors are fluid density, velocity of a fluid particle and stress at a point. Density is a tensor of order 0. It is completely described by $3^0 = 1$ component. The velocity of a particle is a tensor of order 1 ($n = 1$). When expressed in a coordinate system of our choice, velocity has $3^1 = 3$ scalar components. Stress at a point is a tensor of order 2. Thus, when expressed using a coordinate system, it has $3^2 = 9$ scalar components. A tensor of order zero is also called a *scalar*, and that of order 1 is also called a *vector*.

S. S. Sinha, *Fundamentals of Turbulence and Its Modeling*,
https://doi.org/10.1007/978-3-031-94016-3_1

1.1 Expressing a First-Order Tensor Using a Cartesian Coordinate System

Let us have a Cartesian coordinate system $Ox_1(\hat{e}_1)x_2(\hat{e}_2)x_3(\hat{e}_3)$, where O is the origin of the coordinate system (Fig. 1.1). The symbols $\hat{e}_1$, $\hat{e}_2$, and $\hat{e}_3$ are three mutually perpendicular unit vectors. The symbols x_1, x_2 and, x_3 denote the coordinates of an arbitrary location (say P, Fig. 1.1). If $\underline{T}$ is a tensor of order one, we express the tensor in this coordinate system as:

$$\underline{T} = T_1\hat{e}_1 + T_2\hat{e}_2 + T_3\hat{e}_3. \tag{1.1}$$

The symbols T_1, T_2, and T_3 are called the scalar components of the tensor $\underline{T}$ along the unit vectors $\hat{e}_1$, $\hat{e}_2$, and $\hat{e}_3$, respectively. Note that in this book, we represent all tensors (except scalars) using an underlined alphabetical symbol.

1.2 Transformation Rule: First-Order Tensors

Let us now consider another Cartesian coordinate system $Ox_1^{'}(\hat{e}_1^{'})x_2^{'}(\hat{e}_2^{'})x_3^{'}(\hat{e}_3^{'})$ (Fig. 1.2). The same tensor $\underline{T}$ (1.1) can alternatively be expressed in this coordinate system as:

$$\underline{T} = T_1^{'}\hat{e}_1^{'} + T_2^{'}\hat{e}_2^{'} + T_3^{'}\hat{e}_3^{'}. \tag{1.2}$$

In general, $T_1^{'} \neq T_1$, $T_2^{'} \neq T_2$, and $T_3^{'} \neq T_3$. By definition, the tensor $\underline{T}$ must be invariant to the choice of the coordinate system used to express it, the components

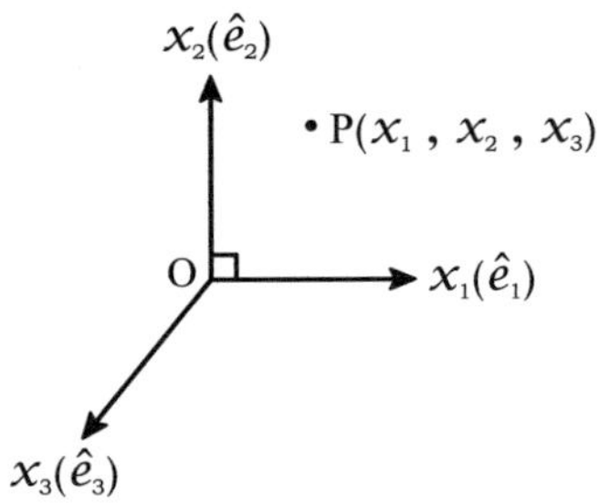

Fig. 1.1 A Cartesian coordinate system. P represents an arbitrary location with coordinates x_1, x_2, x_3

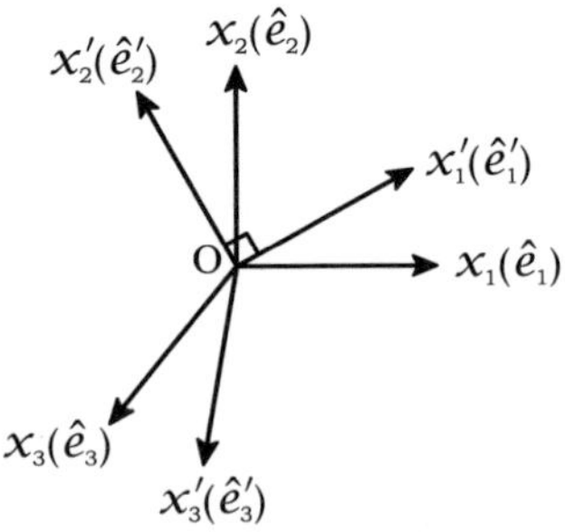

Fig. 1.2 Two different Cartesian coordinate systems

of the tensor in these two coordinate systems must be related. To derive this transformation rule, we first express the unit vectors of the second coordinate system in terms of the unit vectors of the first coordinate system:

$$\begin{aligned}
\hat{e}_1^{'} &= a_{11}\hat{e}_1 + a_{12}\hat{e}_2 + a_{13}\hat{e}_3 = \sum_{i=1}^{3} a_{1i}\hat{e}_i, \\
\hat{e}_2^{'} &= a_{21}\hat{e}_1 + a_{22}\hat{e}_2 + a_{23}\hat{e}_3 = \sum_{i=1}^{3} a_{2i}\hat{e}_i, \\
\hat{e}_3^{'} &= a_{31}\hat{e}_1 + a_{32}\hat{e}_2 + a_{33}\hat{e}_3 = \sum_{i=1}^{3} a_{3i}\hat{e}_i,
\end{aligned} \tag{1.3}$$

where the symbol

$$\sum_{i=1}^{3}$$

represents the sum of three relevant terms that would be generated by allowing the index i to assume values 1, 2, and 3. The coefficients a_{ij}'s appearing on the right-hand side (RHS) of (1.3) are the direction cosines between the unit vectors:

$$a_{ij} = \hat{e}_i^{'} \cdot \hat{e}_j, \tag{1.4}$$

where $i \in \{1, 2, 3\}$ and $j \in \{1, 2, 3\}$. Substituting the equations of unit vectors of the second coordinate system from (1.3) in (1.2) leads to the following equation:

$$\underline{T} = T_1^{'} \sum_{i=1}^{3} a_{1i}\hat{e}_i + T_2^{'} \sum_{i=1}^{3} a_{2i}\hat{e}_i + T_3^{'} \sum_{i=1}^{3} a_{3i}\hat{e}_i. \tag{1.5}$$

Regrouping various terms on the RHS of (1.5) as coefficients of $\hat{e}_1$, $\hat{e}_2$, and $\hat{e}_3$ leads to:

$$\underline{T} = \left\{\sum_{i=1}^{3} T_i^{'} a_{i1}\right\} \hat{e}_1 + \left\{\sum_{i=1}^{3} T_i^{'} a_{i2}\right\} \hat{e}_2 + \left\{\sum_{i=1}^{3} T_i^{'} a_{i3}\right\} \hat{e}_3. \tag{1.6}$$

Comparing (1.6) with (1.1), we arrive at the following set of relationships between the scalar components of the tensor $\underline{T}$ in the two coordinate systems:

$$T_1 = \sum_{i=1}^{3} T_i^{'} a_{i1}, \; T_2 = \sum_{i=1}^{3} T_i^{'} a_{i2}, \; \text{and } T_3 = \sum_{i=1}^{3} T_i^{'} a_{i3}. \tag{1.7}$$

This set of three equations is called the transformation rule of the tensor of order one. These three relationships can be expressed in a more compact manner as:

$$T_j = \sum_{i=1}^{3} T_i^{'} a_{ij}, \tag{1.8}$$

where $j \in \{1, 2, 3\}$ and T_j is the jth scalar component of the tensor $\underline{T}$ in the working coordinate system $Ox_1(\hat{e}_1)x_2(\hat{e}_2)x_3(\hat{e}_3)$.

1.3 Expressing a Second-Order Tensor Using a Cartesian Coordinate System

Now let us consider a second-order tensor $\underline{S}$. In a working coordinate system, it will have 9 scalar components. In the Cartesian coordinate system, $Ox_1(\hat{e}_1)x_2(\hat{e}_2)x_3(\hat{e}_3)$ this tensor is expressed as:

$$\begin{aligned}\underline{S} = S_{11}\hat{e}_1\hat{e}_1 + S_{12}\hat{e}_1\hat{e}_2 + S_{13}\hat{e}_1\hat{e}_3 + S_{21}\hat{e}_2\hat{e}_1 + S_{22}\hat{e}_2\hat{e}_2 + S_{23}\hat{e}_2\hat{e}_3 \\ + S_{31}\hat{e}_3\hat{e}_1 + S_{32}\hat{e}_3\hat{e}_2 + S_{33}\hat{e}_3\hat{e}_3 = \sum_{i=1}^{3}\sum_{j=1}^{3} S_{ij}\hat{e}_i\hat{e}_j,\end{aligned} \tag{1.9}$$

where S_{ij} is the (ij)th scalar component of the tensor $\underline{S}$ in the working coordinate system $Ox_1(\hat{e}_1)x_2(\hat{e}_2)x_3(\hat{e}_3)$. The new mathematical entities $\hat{e}_1\hat{e}_1$, $\hat{e}_1\hat{e}_2$, $\hat{e}_1\hat{e}_3$, $\hat{e}_2\hat{e}_1$, $\hat{e}_2\hat{e}_2$, $\hat{e}_2\hat{e}_3$, $\hat{e}_3\hat{e}_1$, $\hat{e}_3\hat{e}_2$, and $\hat{e}_3\hat{e}_3$ appearing in (1.9) are called *dyads*. Each dyad is an ordered combination of two of the original coordinate system's unit vectors (equivalently, two mutually perpendicular directions). Note that $\hat{e}_1\hat{e}_2 \neq \hat{e}_2\hat{e}_1, \hat{e}_2\hat{e}_3 \neq \hat{e}_3\hat{e}_2$, and$\hat{e}_3\hat{e}_1 \neq \hat{e}_1\hat{e}_3$. The quantity S_{ij} is called the (ij)th component of the tensor $\underline{S}$ along the dyad $\hat{e}_i\hat{e}_j$. Using the summation symbol, S can be expressed equivalently in two ways.

$$\underline{S} = \sum_{i=1}^{3}\sum_{j=1}^{3} S_{ij}\hat{e}_i\hat{e}_j = \sum_{i=1}^{3}\sum_{j=1}^{3} S_{ji}\hat{e}_j\hat{e}_i. \tag{1.10}$$

Equation (1.10) shows that simultaneous changes in the order of the indices of the dyad on the one hand and that of the indices appearing with the scalar components, on the other hand, keep the tensor unchanged.

The *transpose* of a second-order tensor $\underline{S}$ is defined as:

$$\underline{S}^T = \sum_{i=1}^{3}\sum_{j=1}^{3} S_{ji}\hat{e}_i\hat{e}_j = \sum_{i=1}^{3}\sum_{j=1}^{3} S_{ij}\hat{e}_j\hat{e}_i, \tag{1.11}$$

where the symbol $\underline{S}^T$ is the transpose of $\underline{S}$. In general, $\underline{S}^T \neq \underline{S}$.

If a second-order tensor $\underline{S}$ is such that $\underline{S}^T = \underline{S}$, it is called a *symmetric* tensor. In contrast, a second-order tensor $\underline{S}$ is called an *antisymmetric* tensor if $\underline{S}^T = -\underline{S}$. For a symmetric tensor

$$S_{ij} = S_{ji}, \text{ if } i \neq j. \tag{1.12}$$

For an antisymmetric tensor

$$S_{12} = -S_{21}, S_{23} = -S_{32}, \ S_{31} = -S_{13}, \ \text{and } S_{11} = S_{22} = S_{33} = 0. \tag{1.13}$$

Any second-order tensor ($\underline{A}$) can be expressed as the sum of a symmetric tensor ($\underline{A}^{\text{symmetric}}$) and an antisymmetric tensor ($\underline{A}^{\text{antisymmetric}}$):

$$\underline{A} = \underline{A}^{\text{symmetric}} + \underline{A}^{\text{antisymmetric}}, \tag{1.14}$$

where

$$\underline{A}^{\text{symmetric}} = \frac{\underline{A} + \underline{A}^T}{2} \text{ and } \underline{A}^{\text{antisymmetric}} = \frac{\underline{A} - \underline{A}^T}{2}. \tag{1.15}$$

A second-order tensor ($\underline{A}$) can be split into two parts: the isotropic part ($\underline{A}^{\text{isotropic}}$) and the anisotropic part ($\underline{A}^{\text{anisotropic}}$). These parts are defined as

$$\underline{A}^{\text{isotropic}} = \frac{\text{Trace}(\underline{A})}{3}\underline{I} \tag{1.16}$$

and

$$\underline{A}^{\text{anisotropic}} = \underline{A} - \frac{\text{Trace}(\underline{A})}{3}\underline{I} \tag{1.17}$$

The *Trace* of a second-order tensor is defined as

$$\text{Trace}(\underline{A}) = A_{11} + A_{22} + A_{33} \tag{1.18}$$

where the working coordinate system is a Cartesian coordinate system ($(Ox_1(\hat{e}_1)$ $x_2(\hat{e}_2)x_3(\hat{e}_3))$, Fig. 1.1). The tensor $\underline{I}$ represents the *identity tensor of order two*. In the Cartesian coordinate system.

$$\underline{I} = \hat{e}_1\hat{e}_1 + \hat{e}_2\hat{e}_2 + \hat{e}_3\hat{e}_3 \tag{1.19}$$

1.4 Transformation Rule: Second-Order Tensors

We wish to derive the transformation rule of a second-order tensor ($\underline{S}$). Referring to the two coordinate systems of Fig. 1.2, we start with the expression of the tensor $\underline{S}$ in $Ox_1'(\hat{e}_1')x_2'(\hat{e}_2')x_3'(\hat{e}_3')$ coordinate system:

$$\underline{S} = \sum_{i=1}^{3}\sum_{j=1}^{3} S_{ij}'\hat{e}_i'\hat{e}_j'. \tag{1.20}$$

We use (1.3) to substitute the unit vectors of the second coordinate system in terms of the unit vectors of the first coordinate system. Subsequently, we separate the coefficients of the nine dyads, leading to the following expression of $\underline{S}$:

$$\underline{S} = \sum_{p=1}^{3}\sum_{q=1}^{3}\left(\sum_{i=1}^{3}\sum_{j=1}^{3} S_{ij}' a_{ip} a_{jq}\right)\hat{e}_p\hat{e}_q. \tag{1.21}$$

Earlier, we expressed tensor $\underline{S}$ directly in the first coordinate system (1.10), with summation implied over the indices i and j. However, the same tensor be expressed with summations over indices p and q:

$$\underline{S} = \sum_{p=1}^{3}\sum_{q=1}^{3} S_{pq}\hat{e}_p\hat{e}_q. \tag{1.22}$$

Now comparing (1.21) with (1.22) and matching the coefficients of $\hat{e}_p\hat{e}_q$ in the two equations, we arrive at the following expression:

$$S_{pq} = \sum_{i=1}^{3}\sum_{j=1}^{3} S_{ij}' a_{ip} a_{jq}; \text{ where } p, q \in \{1, 2, 3\}. \tag{1.23}$$

The set of nine equations represented by (1.23) is the transformation rule for the second-order tensors.

1.5 Expressing Higher-Order Tensors Using a Cartesian Coordinate System

The manner of expressing the first-order and second-order tensors discussed in previous sections may be extended to higher-order tensors, as well. For an nth-order tensor with 3^n scalar components, we will require 3^n "dyad-like" members constructed with the three mutually perpendicular unit vectors of the chosen coordinate system ($\hat{e}_1$, $\hat{e}_2$, and $\hat{e}_3$). For example, a third-order tensor $\underline{Q}$ is expressed as:

$$\underline{Q} = \sum_{i=1}^{3}\sum_{j=1}^{3}\sum_{k=1}^{3} Q_{ijk}\hat{e}_i\hat{e}_j\hat{e}_k, \tag{1.24}$$

where Q_{ijk} is the (ijk)th scalar component of the tensor $\underline{Q}$ in the given Cartesian coordinate system.

1.6 Einstein's Summation Rule

So far, we have been expressing tensors using one or more summation signs $(\sum)$. While it is indeed a compact way of writing the tensors compared to writing all the 3^n terms explicitly, now onward, we wish to make the expression even more compact. We follow what is called *Einstein's summation rule*. According to this rule, the mere appearance of an index two times in a term by itself implies summation over that index. However, care must be taken that an index must not appear more than two times in any term. Following Einstein's summation rule, the final expressions of various Eqs. (1.5)–(1.24) can be expressed as shown in (1.25)–(1.30).

An index appearing two times in a term is called a *dummy index*. In a term, the choice of an alphabet to represent a pair of dummy indices is not unique. Any symbol can be used as long as a dummy index does not appear more than two times. It is conventional to use the lowercase Latin alphabet to denote these indices. This is illustrated in some examples included in (1.25)–(1.30). In all these examples, every index is repeated exactly two times. The order of the tensor being expressed *must* be inferred based on the number of unit vectors that sequentially appear in the expression. The number of such unit vectors appearing in the sequence defines the order of the tensor. In some cases, the number of pairs of dummy indices may coincidentally match the order of the tensor, but this is not generally true. Example 6 (1.30) represents the dot product of two vectors $\underline{A}$ and $\underline{B}$, which is a tensor of order zero (a scalar, no unit vectors appearing therein). However, the number of pairs of dummy indices is still one.

Example 1.

$$\underline{T} = T_1\hat{e}_1 + T_2\hat{e}_2 + T_3\hat{e}_3 = \sum_{i=1}^{3} T_i\hat{e}_i.$$

The equivalent expression using Einstein's summation rule is

$$\underline{T} = T_i\hat{e}_i = T_j\hat{e}_j. \tag{1.25}$$

Example 2.

$$\hat{e}_1' = \sum_{i=1}^{3} a_{1i}\hat{e}_i.$$

The equivalent expression using Einstein's summation rule is

$$\hat{e}_1' = a_{1i}\hat{e}_i = a_{1k}\hat{e}_k. \tag{1.26}$$

Example 3.

$$\underline{S} = \sum_{i=1}^{3}\sum_{j=1}^{3} S_{ij}\hat{e}_i\hat{e}_j.$$

The equivalent expression using Einstein's summation rule is

$$\underline{S} = S_{ij}\hat{e}_i\hat{e}_j = S_{iq}\hat{e}_i\hat{e}_q = S_{pq}\hat{e}_p\hat{e}_q. \tag{1.27}$$

Example 4.

$$\underline{S}^T = \sum_{i=1}^{3}\sum_{j=1}^{3} S_{ji}\hat{e}_i\hat{e}_j.$$

The equivalent expression using Einstein's summation rule is

$$\underline{S}^T = S_{ji}\hat{e}_i\hat{e}_j. \tag{1.28}$$

Example 5.

$$\underline{Q} = \sum_{i=1}^{3}\sum_{j=1}^{3}\sum_{k=1}^{3} Q_{ijk}\hat{e}_i\hat{e}_j\hat{e}_k.$$

The equivalent expression using Einstein's summation rule is

$$\underline{Q} = Q_{ijk}\hat{e}_i\hat{e}_j\hat{e}_k = Q_{ijr}\hat{e}_i\hat{e}_j\hat{e}_r = Q_{qjr}\hat{e}_q\hat{e}_j\hat{e}_r. \tag{1.29}$$

Example 6.

$$\phi = A_1B_1 + A_2B_2 + A_3B_3 = \sum_{j=1}^{3} A_jB_j.$$

The equivalent expression using Einstein's summation rule is

$$\phi = A_jB_j. \tag{1.30}$$

Einstein's summation rule can also be used to express individual scalar components of a tensor. In (1.31) and (1.32), we include such examples (Examples 7 and 8). Note that there are already one or more pairs of dummy indices in all these expressions. Further, in each expression, there is one or more non-repeated index. Such a non-repeated index is called a *free index*. The number of free indices in an expression always matches the order of the original tensor to which this scalar component belongs. These free indices may be assigned values 1, 2 or 3 to express various components of the tensor.

Example 7.

$$T_j = \sum_{i=1}^{3} T_i^{'} a_{ij}.$$

The equivalent expression using Einstein's summation rule is

$$T_j = T_i^{'} a_{ij}. \tag{1.31}$$

Example 8.

$$S_{pq} = \sum_{i=1}^{3}\sum_{j=1}^{3} S'_{ij} a_{ip} a_{jq}.$$

The equivalent expression using Einstein's summation rule is

$$S_{pq} = S'_{ij} a_{ip} a_{jq}. \tag{1.32}$$

In an equation, if there are free indices, all terms of the equation must have the same number of free indices, and further, those free indices must be identical in all the terms on either side of the equation. For example, in the equation

$$S_{pq} = S'_{ij} a_{ip} a_{jq}, \tag{1.33}$$

the term on the left-hand side (LHS) and that on the RHS have exactly two free indices. Moreover, these free indices are p and q in each term. This manner of expressing tensors in a Cartesian coordinate system using free and dummy indices is called the *index notation*.

1.7 Tensor Operations

In this section, we define some tensor operations relevant to our study of fluid mechanics.

1.7.1 Dot Product of Two Tensors

Let us consider two tensors $\underline{T}$ (of order $t \geq 1$) and $\underline{N}$ (of order $n \geq 1$). We define two types of dot product between two tensors: $\underline{T} \cdot \underline{N}$ and $\underline{N} \cdot \underline{T}$. In general, these two dot products result in two different tensors. However, in the special case when both these tensors are of order one, then, $\underline{T} \cdot \underline{N} = \underline{N} \cdot \underline{T}$. The execution of a dot product always results in a tensor with an order that is two less than the sum of the orders of the two participating tensors.

To illustrate the algebraic implementation of a dot product, we consider the dot product of a second-order tensor with a first-order tensor. We perform this illustration using the Cartesian coordinate system of Fig. 1.1 as our working coordinate system. We first express the two participating tensors in a Cartesian coordinate system and subsequently simplify the algebra as much as possible.

$$\begin{aligned}\underline{T} \cdot \underline{N} = & \left[T_{11}\hat{e}_1\hat{e}_1 + T_{12}\hat{e}_1\hat{e}_2 + T_{13}\hat{e}_1\hat{e}_3 + T_{21}\hat{e}_2\hat{e}_1 + T_{22}\hat{e}_2\hat{e}_2 + T_{23}\hat{e}_2\hat{e}_3\right. \\ & \left. + \; T_{31}\hat{e}_3\hat{e}_1 + T_{32}\hat{e}_3\hat{e}_2 + T_{33}\hat{e}_3\hat{e}_3\right] \cdot \left[N_1\hat{e}_1 + N_2\hat{e}_2 + N_3\hat{e}_3\right].\end{aligned} \tag{1.34}$$

The right-hand side of Eq. (1.35) results into 27 individual terms.

$$\begin{aligned}
\underline{T}\cdot\underline{N} = &\ T_{11}\hat{e}_1\hat{e}_1\cdot N_1\hat{e}_1 + T_{12}\hat{e}_1\hat{e}_2\cdot N_1\hat{e}_1 + T_{13}\hat{e}_1\hat{e}_3\cdot N_1\hat{e}_1\\
+ &\ T_{21}\hat{e}_2\hat{e}_1\cdot N_1\hat{e}_1 + T_{22}\hat{e}_2\hat{e}_2\cdot N_1\hat{e}_1 + T_{23}\hat{e}_2\hat{e}_3\cdot N_1\hat{e}_1\\
+ &\ T_{31}\hat{e}_3\hat{e}_1\cdot N_1\hat{e}_1 + T_{32}\hat{e}_3\hat{e}_2\cdot N_1\hat{e}_1 + T_{33}\hat{e}_3\hat{e}_3\cdot N_1\hat{e}_1\\
+ &\ T_{11}\hat{e}_1\hat{e}_1\cdot N_2\hat{e}_2 + T_{12}\hat{e}_1\hat{e}_2\cdot N_2\hat{e}_2 + T_{13}\hat{e}_1\hat{e}_3\cdot N_2\hat{e}_2\\
+ &\ T_{21}\hat{e}_2\hat{e}_1\cdot N_2\hat{e}_2 + T_{22}\hat{e}_2\hat{e}_2\cdot N_2\hat{e}_2 + T_{23}\hat{e}_2\hat{e}_3\cdot N_2\hat{e}_2\\
+ &\ T_{31}\hat{e}_3\hat{e}_1\cdot N_2\hat{e}_2 + T_{32}\hat{e}_3\hat{e}_2\cdot N_2\hat{e}_2 + T_{33}\hat{e}_3\hat{e}_3\cdot N_2\hat{e}_2\\
+ &\ T_{11}\hat{e}_1\hat{e}_1\cdot N_3\hat{e}_3 + T_{12}\hat{e}_1\hat{e}_2\cdot N_3\hat{e}_3 + T_{13}\hat{e}_1\hat{e}_3\cdot N_3\hat{e}_3\\
+ &\ T_{21}\hat{e}_2\hat{e}_1\cdot N_3\hat{e}_3 + T_{22}\hat{e}_2\hat{e}_2\cdot N_3\hat{e}_3 + T_{23}\hat{e}_2\hat{e}_3\cdot N_3\hat{e}_3\\
+ &\ T_{31}\hat{e}_3\hat{e}_1\cdot N_3\hat{e}_3 + T_{32}\hat{e}_3\hat{e}_2\cdot N_3\hat{e}_3 + T_{33}\hat{e}_3\hat{e}_3\cdot N_3\hat{e}_3.
\end{aligned} \tag{1.35}$$

The rule by which the dot product operates is that in an expression like $T_{12}\hat{e}_1\hat{e}_2\cdot N_1\hat{e}_1$ of (1.35), *the rightmost unit vector from the expression appearing on the left side of the dot operator should dot with the leftmost unit vector from the expression appearing on the right-hand side of the dot operator*. In other words, *the "nearest" two unit vectors must dot with each other*. Thus,

$$T_{12}\hat{e}_1\hat{e}_2\cdot N_1\hat{e}_1 = T_{12}N_1\hat{e}_1\left(\hat{e}_2\cdot\hat{e}_1\right) = T_{12}N_1\hat{e}_1\,(0) = 0. \tag{1.36}$$

Since the three unit vectors ($\hat{e}_1$, $\hat{e}_2$, and $\hat{e}_3$) are mutually perpendicular to each other, out of the 27 individual dot products of (1.35), only those survive which involve the dot products of the same unit vectors ($\hat{e}_1\cdot\hat{e}_1$, $\hat{e}_2\cdot\hat{e}_2$ and $\hat{e}_3\cdot\hat{e}_3$). Accordingly, there are nine such surviving terms.

$$\begin{aligned}
\underline{T}\cdot\underline{N} = &\ T_{11}N_1\hat{e}_1 + T_{12}N_2\hat{e}_1 + T_{13}N_3\hat{e}_1 + T_{21}N_1\hat{e}_2\\
&+ T_{22}N_2\hat{e}_2 + T_{23}N_3\hat{e}_2 + T_{31}N_1\hat{e}_3 + T_{32}N_2\hat{e}_3 + T_{33}N_3\hat{e}_3.
\end{aligned} \tag{1.37}$$

Using Einstein's summation rule, this dot product is written as:

$$\underline{T}\cdot\underline{N} = T_{ij}N_j\hat{e}_i. \tag{1.38}$$

Now let us consider the dot product $\underline{N}\cdot\underline{T}$. This can be expressed as:

$$\begin{aligned}
\underline{N}\cdot\underline{T} = &\left[N_1\hat{e}_1 + N_2\hat{e}_2 + N_3\hat{e}_3\right]\cdot\left[T_{11}\hat{e}_1\hat{e}_1 + T_{12}\hat{e}_1\hat{e}_2 + T_{13}\hat{e}_1\hat{e}_3\right.\\
&\left. T_{21}\hat{e}_2\hat{e}_1 + T_{22}\hat{e}_2\hat{e}_2 + T_{23}\hat{e}_2\hat{e}_3 + T_{31}\hat{e}_3\hat{e}_1 + T_{32}\hat{e}_3\hat{e}_2 + T_{33}\hat{e}_3\hat{e}_3\right].
\end{aligned} \tag{1.39}$$

Following the same rule (the nearest two unit vectors dot with each other), we arrive at:

$$\begin{aligned}
\underline{N}\cdot\underline{T} = &\ N_1T_{11}\hat{e}_1 + N_1T_{12}\hat{e}_2 + N_1T_{13}\hat{e}_3 + N_2T_{21}\hat{e}_1 + N_2T_{22}\hat{e}_2\\
&+ N_2T_{23}\hat{e}_3 + N_3T_{31}\hat{e}_1 + N_3T_{32}\hat{e}_2 + N_3T_{33}\hat{e}_3.
\end{aligned} \tag{1.40}$$

Using Einstein's summation rule (1.40) is written as:

$$\underline{N} \cdot \underline{T} = N_j T_{ji} \hat{e}_i = T_{ji} N_j \hat{e}_i. \tag{1.41}$$

In the expressions of both $\underline{T} \cdot \underline{N}$ and $\underline{N} \cdot \underline{T}$ we have only one unit vector appearing on the RHS of (1.38) and (1.41). This means that both $\underline{T} \cdot \underline{N}$ and $\underline{N} \cdot \underline{T}$ are vectors (tensors of order one). Now, for the two vectors to be identical, we must compare the ith components of the two vectors. The ith component of a vector is the coefficient of $\hat{e}_i$ in its expression (1.38 or 1.41). The ith components of $\underline{T} \cdot \underline{N}$ in (1.38) and that of $\underline{N} \cdot \underline{T}$ in (1.41) are $T_{ij} N_j$ and $T_{ji} N_j$, respectively. Since $T_{ij} N_j (= T_{i1} N_1 + T_{i2} N_2 + T_{i3} N_3) \neq T_{ji} N_j (= T_{1i} N_1 + T_{2i} N_2 + T_{3i} N_3)$, we conclude that $\underline{T} \cdot \underline{N} \neq \underline{N} \cdot \underline{T}$.

At this point, we introduce a new symbol called the *Kronecker delta* (δ). This symbol has two indices as subscripts, with which we define

$$\delta_{ij} = \hat{e}_i \cdot \hat{e}_j, \tag{1.42}$$

where $\hat{e}_i$ and $\hat{e}_j$ are the ith and jth unit vectors of our Cartesian coordinate system. Clearly, (1.42) leads to:

$$\delta_{ij} = \begin{cases} 1 & \text{if } (i, j) = (1, 1), \text{ or } (2, 2), \text{ or } (3, 3) \\ 0 & \text{if } i \neq j. \end{cases} \tag{1.43}$$

Note that following the definition of δ_{ij} and simultaneously using Einstein's summation rule

$$\delta_{ii} = \delta_{11} + \delta_{22} + \delta_{33} = 3. \tag{1.44}$$

The specific purpose of introducing the Kronecker delta symbol here is to symbolically represent the dot product between the two unit vectors of the working Cartesian coordinate system. With this definition, one can avoid dealing with the expanded forms of various tensors (like what we had to do in (1.35) and (1.39)). We illustrate this by revisiting the dot product of (1.34).

$$\underline{T} \cdot \underline{N} = \left(T_{ij} \hat{e}_i \hat{e}_j\right) \cdot \left(N_p \hat{e}_p\right) = T_{ij} N_p \hat{e}_i \hat{e}_j \cdot \hat{e}_p = T_{ij} N_p \hat{e}_i \delta_{jp}. \tag{1.45}$$

Once a Kronecker delta symbol appears in a term (δ_{jp} in this particular example), we perform the following two simplifying steps: (1) remove the Kronecker delta symbol, and (2) replace either the remaining j by p or replace the remaining p by j in the term. By performing these two steps, we achieve (i) the removal of all those dot products wherein two different unit vectors participate and consequently vanish and (ii) retaining all those dot products wherein identical unit vectors participate. Employing these two steps, (1.45) readily simplifies as:

$$\underline{T} \cdot \underline{N} = T_{ij} N_p \hat{e}_i \delta_{jp} = T_{ij} N_j \hat{e}_i. \tag{1.46}$$

Alternatively,

$$\underline{T} \cdot \underline{N} = T_{ij} N_p \hat{e}_i \delta_{jp} = T_{ip} N_p \hat{e}_i, \tag{1.47}$$

which leads to the same outcome as what we obtained in (1.38).

1.7.2 Double Dot Product of Two Tensors

The double-dot product is defined for two tensors $\underline{T}$ and $\underline{N}$ of orders t and n, if $t \geq 2$ and $n \geq 2$. The resulting tensor is of order $t + n - 4$. The double dot product operation is defined by performing two successive dot products between the right-most unit vector of the tensor on the left and the left-most unit vector of the tensor on the right. We illustrate the double-dot operation using an example wherein two second-order tensors are participating:

$$
\begin{aligned}
\underline{T} : \underline{N} = & \left(T_{ij}\hat{e}_i\hat{e}_j\right) : \left(N_{pq}\hat{e}_p\hat{e}_q\right) = T_{ij}N_{pq}\hat{e}_i\hat{e}_j : \hat{e}_p\hat{e}_q, \\
= & \; T_{ij}N_{pq}\delta_{jp}\hat{e}_i \cdot \hat{e}_q = T_{ij}N_{jq}\hat{e}_i \cdot \hat{e}_q = T_{ij}N_{jq}\delta_{iq}, \\
= & \; T_{ij}N_{ji},
\end{aligned} \tag{1.48}
$$

which is a scalar (because no unit vector appears in the final expression).

1.7.3 Cross Product of Two Vectors

The cross-product is defined for two vectors. The resulting quantity is also a vector. The cross product between vector $\underline{A}$ and $\underline{B}$ is defined as

$$
\underline{A} \times \underline{B} = \varepsilon_{ijk}A_iB_j\hat{e}_k, \tag{1.49}
$$

where the symbol ε is called the *permutation symbol* such that,

$$
\varepsilon_{ijk} = \begin{cases} +1 & \text{if } i \neq j \neq k \text{ and } (i, j, k) \text{ follows a cyclic order} \\ & (1, 2, 3) \text{ or } (2, 3, 1) \text{ or } (3, 1, 2), \\ -1 & \text{if } i \neq j \neq k \text{ and } (i, j, k) \text{ follows the reversed cyclic order} \\ & (1, 3, 2) \text{ or } (3, 2, 1) \text{ or } (2, 1, 3), \\ 0 & \text{otherwise.} \end{cases} \tag{1.50}
$$

Based on this definition, it follows that only when $i \neq j \neq k$, $\varepsilon_{ijk} \neq 0$. It can be verified that in case $i \neq j \neq k$, the swapping of the positions of a pair of indices changes the sign of ε_{ijk}

$$
\varepsilon_{ijk} = -\varepsilon_{jik} = -(-\varepsilon_{jki}) = \varepsilon_{jki}. \tag{1.51}
$$

It can be verified that the permutation symbol (1.50) leads to the following relationships

$$
\begin{aligned}
\hat{e}_1 \times \hat{e}_2 &= \hat{e}_3, & \hat{e}_2 \times \hat{e}_3 &= \hat{e}_1, & \hat{e}_3 \times \hat{e}_1 &= \hat{e}_2, \\
\hat{e}_2 \times \hat{e}_1 &= -\hat{e}_3, & \hat{e}_3 \times \hat{e}_2 &= -\hat{e}_1, & \hat{e}_1 \times \hat{e}_3 &= -\hat{e}_2.
\end{aligned} \tag{1.52}
$$

The RHS of (1.49) does lead to the following familiar expanded form of the cross product of two vectors

$$
\underline{A} \times \underline{B} = (A_2B_3 - A_3B_2)\,\hat{e}_1 + (A_3B_1 - A_1B_3)\,\hat{e}_2 + (A_1B_2 - A_2B_1)\,\hat{e}_3. \tag{1.53}
$$

1.8 The ϵ-δ Identity

It can be verified that the following relationship exists between the permutation symbol (ε) and the Kronecker delta (δ) symbols.

$$\varepsilon_{ijk}\varepsilon_{imn} = \delta_{jm}\delta_{kn} - \delta_{jn}\delta_{km}. \tag{1.54}$$

Equation (1.54) is called the ϵ-δ *identity* and often proves useful while performing algebraic manipulations of expressions involving multiple cross products. Note that the first index of the two permutation symbols on the LHS of (1.54) are identical.

1.9 Spatial Derivatives of Tensors

In later sections, when we derive the governing equations of fluid motion, we come across spatial derivatives of various kinematic and force-related quantities. Thus, we must define an operator ($\underline{\nabla}$) with the help of which various derivatives of space-dependent tensors can be expressed and algebraically manipulated. We refer to this operator as the *nabla* operator. Using the Cartesian coordinate system of Fig. 1.1, $\underline{\nabla}$ is expressed as:

$$\underline{\nabla} = \hat{e}_1\frac{\partial}{\partial x_1} + \hat{e}_2\frac{\partial}{\partial x_2} + \hat{e}_3\frac{\partial}{\partial x_3}, \text{ or } \underline{\nabla} = \hat{e}_m\frac{\partial}{\partial x_m}, \tag{1.55}$$

where $\partial/\partial x_m$ is the partial derivative operator with respect to the spatial coordinate x_m where $m \in \{1, 2, 3\}$. In the context of this book, there are five specific operations of the nabla operator which we need to understand. These are the *gradient of a tensor*, the *divergence of a tensor*, the *advection operator*, the *curl of a vector* and the *Laplacian of a tensor*,

The gradient of a tensor of order t (where $t \geq 0$) results in a tensor of order $t + 1$. We illustrate this operation using an example where $t = 2$.

$$\underline{\nabla T} = \left(\hat{e}_m\frac{\partial}{\partial x_m}\right)\left(T_{ij}\hat{e}_i\hat{e}_j\right) = \hat{e}_m\frac{\partial T_{ij}}{\partial x_m}\hat{e}_i\hat{e}_j = \frac{\partial T_{ij}}{\partial x_m}\hat{e}_m\hat{e}_i\hat{e}_j. \tag{1.56}$$

Since the unit vectors $\hat{e}_1$, $\hat{e}_2$, and $\hat{e}_3$ do not depend on the spatial coordinates (x_1, x_2, and x_3), the spatial derivatives of these unit vectors do not appear in (1.56). The final expression in (1.56) has an ordered sequence of three unit vectors, which clearly shows that the resulting tensor is a third-order tensor.

The divergence of a tensor of order t is defined if $t \geq 1$. The resulting tensor is of order $(t - 1)$. We illustrate this operation using an example where $t = 2$.

$$\begin{aligned}\underline{\nabla}\cdot\underline{T} = & \left(\hat{e}_m\frac{\partial}{\partial x_m}\right)\cdot\left(T_{ij}\hat{e}_i\hat{e}_j\right) = \frac{\partial T_{ij}}{\partial x_m}\hat{e}_m\cdot\hat{e}_i\hat{e}_j = \frac{\partial T_{ij}}{\partial x_m}\delta_{mi}\hat{e}_j,\\ = & \frac{\partial T_{ij}}{\partial x_i}\hat{e}_j,\end{aligned} \tag{1.57}$$

which is a first-order tensor (only one unit vector appears on the RHS of (1.57)).

The Laplacian operator (∇^2) is defined as:

$$\nabla^2 = \left(\underline{\nabla} \cdot \underline{\nabla}\right). \tag{1.58}$$

The Laplacian operator can act on a tensor of any order. The resulting tensor has the same order as the tensor on which the operator acted. In the Cartesian coordinate system, the Laplacian operator is expressed as:

$$\begin{aligned} \nabla^2 = \left(\underline{\nabla} \cdot \underline{\nabla}\right) &= \hat{e}_m \frac{\partial}{\partial x_m} \cdot \hat{e}_n \frac{\partial}{\partial x_n} = \hat{e}_m \cdot \hat{e}_n \frac{\partial^2}{\partial x_m \partial x_n} = \delta_{mn} \frac{\partial^2}{\partial x_m \partial x_n}, \\ &= \frac{\partial^2}{\partial x_m \partial x_m}. \end{aligned} \tag{1.59}$$

We illustrate the effect of this operator on a second-order tensor ($\underline{T}$):

$$\nabla^2 \underline{T} = \left(\underline{\nabla} \cdot \underline{\nabla}\right) \underline{T} = \frac{\partial^2}{\partial x_m \partial x_m} T_{ij} \hat{e}_i \hat{e}_j = \frac{\partial^2 T_{ij}}{\partial x_m \partial x_m} \hat{e}_i \hat{e}_j. \tag{1.60}$$

The right-most expression in (1.60) has an ordered pair of unit vectors ($\hat{e}_i \hat{e}_j$), which implies the resulting tensor is a second-order tensor, like $\underline{T}$, itself. However, $\nabla^2 \underline{T} \neq \underline{T}$.

The advection operator is defined as:

$$\left(\underline{V} \cdot \underline{\nabla}\right), \tag{1.61}$$

where $\underline{V}$ is a vector. The advection operator can act on a tensor of any order. The resulting tensor is of the same order as the original tensor on which the advection operator acts. Let us consider an example wherein the advection operator acts on a vector $\underline{B}$.

$$\begin{aligned} \left(\underline{V} \cdot \underline{\nabla}\right) \underline{B} &= \left(V_m \hat{e}_m \cdot \hat{e}_n \frac{\partial}{\partial x_n}\right) B_q \hat{e}_q = V_m \delta_{mn} \frac{\partial}{\partial x_n} B_q \hat{e}_q, \\ &= \left(V_n \frac{\partial}{\partial x_n}\right) B_q \hat{e_q} = V_n \frac{\partial B_q}{\partial x_n} \hat{e_q}. \end{aligned} \tag{1.62}$$

The parentheses appearing on the LHS of (1.62) imply that the dot product operation must first be performed between $\underline{V}$ and $\underline{\nabla}$, and subsequently, the resulting operator acts on $\underline{B}$.

The curl of a vector is defined as the cross product between the *nabla* operator and a vector:

$$\underline{\nabla} \times \underline{V}. \tag{1.63}$$

Expressing both the nabla operator and the vector $\underline{V}$ using our Cartesian coordinate system and (1.49), (1.63) is expressed as

$$\underline{\nabla} \times \underline{V} = \hat{e}_m \frac{\partial}{\partial x_m} \times V_p \hat{e}_p = \hat{e}_m \times \hat{e}_p \frac{\partial V_p}{\partial x_m} = \varepsilon_{mpn} \frac{\partial V_p}{\partial x_m} \hat{e}_n. \tag{1.64}$$

1.10 Index Notation and Tensor Identities

In previous sections, we introduced the index notation, primarily to enable us to express a tensor in the Cartesian coordinate system in a compact manner. Additionally, the index notation also proves useful in demonstrating the proofs of various tensor identities. Since tensors themselves remain invariant to the choice of the working coordinate system, so are the tensor identities. Thus, it is adequate to prove a tensor identity using any one working coordinate system of our choice. Our Cartesian coordinate system, for which we use the index notation for brevity, is indeed an apt choice to demonstrate the proofs of tensor identities. For this purpose, we first use index notation to express one side (the LHS or the RHS) of the identity in the Cartesian coordinate system. Subsequently, various rules of the index notation, along with the relevant properties of the permutation symbol and the Kronecker delta symbol, are employed to simplify (and sometimes expand) the algebraic terms. Finally, these modified algebraic terms are converted back to the form which is independent of the choice of the working coordinate system. This process is illustrated using the following examples.

Example 1. Prove that $\underline{\nabla} \times \left(\underline{\nabla}\phi\right) = 0$.

$$\underline{\nabla} \times \left(\underline{\nabla}\phi\right) = \underline{\nabla} \times \left(\hat{e}_m \frac{\partial \phi}{\partial x_m}\right) = \varepsilon_{imk} \frac{\partial}{\partial x_i}\left(\frac{\partial \phi}{\partial x_m}\right)\hat{e}_k = \varepsilon_{imk} \frac{\partial^2 \phi}{\partial x_i \partial x_m}\hat{e}_k.$$

Now, we carefully change all i's to m's and all m's to i's in the last expression. The resulting tensor must remain unchanged because both i and m are dummy indices. Thus,

$$\varepsilon_{imk} \frac{\partial^2 \phi}{\partial x_i \partial x_m}\hat{e}_k = \varepsilon_{mik} \frac{\partial^2 \phi}{\partial x_m \partial x_i}\hat{e}_k.$$

However, merely interchanging the positions of i and k in the permutation symbol on the RHS must reverse the sign of the tensor

$$\varepsilon_{imk} \frac{\partial^2 \phi}{\partial x_i \partial x_m}\hat{e}_k = -\varepsilon_{imk} \frac{\partial^2 \phi}{\partial x_m \partial x_i}\hat{e}_k. \tag{1.65}$$

Since $\frac{\partial^2 \phi}{\partial x_m \partial x_i} = \frac{\partial^2 \phi}{\partial x_i \partial x_m}$, the LHS and the RHS of (1.65) are identical except for the negative sign. For this to be true, both LHS and RHS must be zero. Thus,

$$\varepsilon_{imk} \frac{\partial^2 \phi}{\partial x_i \partial x_m}\hat{e}_k = 0 \Rightarrow \underline{\nabla} \times \left(\underline{\nabla}\phi\right) = 0. \tag{1.66}$$

Example 2. Prove that $\underline{\nabla} \cdot (\underline{\nabla} \times \underline{V}) = 0$.

$$\begin{aligned}
\underline{\nabla} \cdot (\underline{\nabla} \times \underline{V}) &= \hat{e}_m \frac{\partial}{\partial x_m} \cdot \left(\hat{e}_p \frac{\partial}{\partial x_p} \times V_q \hat{e}_q \right) \\
&= \hat{e}_m \frac{\partial}{\partial x_m} \cdot \left(\varepsilon_{pqr} \frac{\partial V_q}{\partial x_p} \hat{e}_r \right) = \varepsilon_{pqr} \frac{\partial^2 V_q}{\partial x_m \partial x_p} \hat{e}_m \cdot \hat{e}_r, \\
&= \varepsilon_{pqr} \frac{\partial^2 V_q}{\partial x_m \partial x_p} \delta_{mr} = \varepsilon_{pqm} \frac{\partial^2 V_q}{\partial x_m \partial x_p}. \qquad (1.67)
\end{aligned}$$

Now, we carefully change all p's to m's and all m's to p's in the last expression. The resulting tensor must remain unchanged. Thus,

$$\varepsilon_{pqm} \frac{\partial^2 V_q}{\partial x_m \partial x_p} = \varepsilon_{mqp} \frac{\partial^2 V_q}{\partial x_p \partial x_m}. \qquad (1.68)$$

However, merely interchanging the positions of m and p in the permutation symbol on the RHS must reverse the sign of the tensor leading to

$$\varepsilon_{pqm} \frac{\partial^2 V_q}{\partial x_m \partial x_p} = -\varepsilon_{pqm} \frac{\partial^2 V_q}{\partial x_p \partial x_m}. \qquad (1.69)$$

We observe that the LHS and the RHS of (1.69) are identical except for the negative sign. For this to be true, both LHS and RHS must be zero.

$$\varepsilon_{pqm} \frac{\partial^2 V_q}{\partial x_p \partial x_m} = 0 \Rightarrow \underline{\nabla} \cdot (\underline{\nabla} \times \underline{V}) = 0. \qquad (1.70)$$

Example 3. Prove that

$$\underline{\nabla} \times (\underline{\omega} \times \underline{V}) = (\underline{V} \cdot \underline{\nabla})\, \underline{\omega} - (\underline{\nabla} \cdot \underline{\omega})\, \underline{V} + (\underline{\nabla} \cdot \underline{V})\, \underline{\omega} - (\underline{\omega} \cdot \underline{\nabla})\, \underline{V}.$$

$$\begin{aligned}
\underline{\nabla} \times (\underline{\omega} \times \underline{V}) &= \underline{\nabla} \times \left(\varepsilon_{ijk} \omega_i V_j \hat{e}_k \right) = \epsilon_{pkr} \frac{\partial}{\partial x_p} \left(\varepsilon_{ijk} \omega_i V_j \right) \hat{e}_r, \\
&= \varepsilon_{pkr} \varepsilon_{ijk} \frac{\partial \omega_i}{\partial x_p} V_j \hat{e}_r + \varepsilon_{pkr} \varepsilon_{ijk} \frac{\partial V_j}{\partial x_p} \omega_i \hat{e}_r. \qquad (1.71)
\end{aligned}$$

Both the terms in the RHS of the last expression in (1.71) have two permutation symbols with one common index. The $\epsilon - \delta$ identity (1.54) can be applied here. However, before we do so, we must re-arrange the indices of both the permutation symbols such that the common index appears as the first index of each of these symbols.

$$
\begin{aligned}
\varepsilon_{pkr}\varepsilon_{ijk}\frac{\partial\omega_i}{\partial x_p}V_j\hat{e}_r + \varepsilon_{pkr}\varepsilon_{ijk}\frac{\partial V_j}{\partial x_p}\omega_i\hat{e}_r
&= (-\varepsilon_{kpr})(-\varepsilon_{kji})\frac{\partial\omega_i}{\partial x_p}V_j\hat{e}_r + (-\varepsilon_{kpr})(-\varepsilon_{kji})\frac{\partial V_j}{\partial x_p}\omega_i\hat{e}_r,\\
&= \varepsilon_{kpr}\varepsilon_{kji}\frac{\partial\omega_i}{\partial x_p}V_j\hat{e}_r + \varepsilon_{kpr}\varepsilon_{kji}\frac{\partial V_j}{\partial x_p}\omega_i\hat{e}_r,\\
&= \left[\delta_{pj}\delta_{ri}\frac{\partial\omega_i}{\partial x_p}V_j\hat{e}_r - \delta_{pi}\delta_{rj}\frac{\partial\omega_i}{\partial x_p}V_j\hat{e}_r\right]\\
&\quad + \left[\delta_{pj}\delta_{ri}\frac{\partial V_j}{\partial x_p}\omega_i\hat{e}_r - \delta_{pi}\delta_{rj}\frac{\partial V_j}{\partial x_p}\omega_i\hat{e}_r\right],\\
&= \left[V_p\frac{\partial\omega_r}{\partial x_p}\hat{e}_r - \frac{\partial\omega_p}{\partial x_p}V_r\hat{e}_r\right] + \left[\frac{\partial V_p}{\partial x_p}\omega_r\hat{e}_r - \omega_p\frac{\partial V_r}{\partial x_p}\hat{e}_r\right],\\
&= \left[\left(\underline{V}\cdot\underline{\nabla}\right)\underline{\omega} - \left(\underline{\nabla}\cdot\underline{\omega}\right)\underline{V}\right] + \left[\left(\underline{\nabla}\cdot\underline{V}\right)\underline{\omega} - \left(\underline{\omega}\cdot\underline{\nabla}\right)\underline{V}\right],
\end{aligned}
$$

$$
\Rightarrow \underline{\nabla}\times\left(\underline{\omega}\times\underline{V}\right) = \left(\underline{V}\cdot\underline{\nabla}\right)\underline{\omega} - \left(\underline{\nabla}\cdot\underline{\omega}\right)\underline{V} + \left(\underline{\nabla}\cdot\underline{V}\right)\underline{\omega} - \left(\underline{\omega}\cdot\underline{\nabla}\right)\underline{V}. \tag{1.72}
$$

Example 4. Prove that $\underline{V}\times\left(\underline{\nabla}\times\underline{V}\right) = \frac{1}{2}\underline{\nabla}\left(\underline{V}\cdot\underline{V}\right) - \left(\underline{V}\cdot\underline{\nabla}\right)\underline{V}$.

$$
\begin{aligned}
\underline{V}\times\left(\underline{\nabla}\times\underline{V}\right) &= V_p\hat{e}_p\times\left(\hat{e}_i\frac{\partial}{\partial x_i}\times V_j\hat{e}_j\right) = V_p\hat{e}_p\times\left(\varepsilon_{ijk}\frac{\partial V_j}{\partial x_i}\hat{e}_k\right),\\
&= \varepsilon_{pkr}V_p\varepsilon_{ijk}\frac{\partial V_j}{\partial x_i}\hat{e}_r = \left(-\varepsilon_{kpr}\right)V_p\left(-\varepsilon_{kji}\right)\frac{\partial V_j}{\partial x_i}\hat{e}_r,\\
&= \varepsilon_{kpr}\varepsilon_{kji}V_p\frac{\partial V_j}{\partial x_i}\hat{e}_r = \left(\delta_{pj}\delta_{ri} - \delta_{pi}\delta_{rj}\right)V_p\frac{\partial V_j}{\partial x_i}\hat{e}_r,\\
&= \delta_{pj}\delta_{ri}V_p\frac{\partial V_j}{\partial x_i}\hat{e}_r - \delta_{pi}\delta_{rj}V_p\frac{\partial V_j}{\partial x_i}\hat{e}_r,\\
&= \delta_{ri}V_j\frac{\partial V_j}{\partial x_i}\hat{e}_r - \delta_{rj}V_i\frac{\partial V_j}{\partial x_i}\hat{e}_r = V_j\frac{\partial V_j}{\partial x_i}\hat{e}_i - V_i\frac{\partial V_j}{\partial x_i}\hat{e}_j,\\
&= \hat{e}_i\frac{1}{2}\frac{\partial\left(V_jV_j\right)}{\partial x_i} - V_i\frac{\partial V_j}{\partial x_i}\hat{e}_j,\\
&= \underline{\nabla}\left[\frac{\left(\underline{V}\cdot\underline{V}\right)}{2}\right] - \left(\underline{V}\cdot\underline{\nabla}\right)\underline{V},
\end{aligned}
$$

$$
\Rightarrow \underline{V}\times\left(\underline{\nabla}\times\underline{V}\right) = \frac{\underline{\nabla}\left(\underline{V}\cdot\underline{V}\right)}{2} - \left(\underline{V}\cdot\underline{\nabla}\right)\underline{V}. \tag{1.73}
$$

Example 5. Prove that $\left[(\underline{\nabla}\phi)\cdot\underline{\nabla}\right]\underline{\nabla}\phi = \frac{1}{2}\underline{\nabla}\left[(\underline{\nabla}\phi)\cdot(\underline{\nabla}\phi)\right]$.

$$\begin{aligned}
\left[(\underline{\nabla}\phi)\cdot\underline{\nabla}\right]\underline{\nabla}\phi &= \left[\left(\hat{e}_m\frac{\partial}{\partial x_m}\phi\right)\cdot\hat{e}_n\frac{\partial}{\partial x_n}\right]\hat{e}_p\frac{\partial\phi}{\partial x_p},\\
&= \hat{e}_m\cdot\hat{e}_n\frac{\partial\phi}{\partial x_m}\frac{\partial}{\partial x_n}\left(\hat{e}_p\frac{\partial\phi}{\partial x_p}\right) = \delta_{mn}\frac{\partial\phi}{\partial x_m}\frac{\partial^2\phi}{\partial x_n\partial x_p}\hat{e}_p,\\
&= \hat{e}_p\frac{\partial\phi}{\partial x_m}\frac{\partial^2\phi}{\partial x_m\partial x_p} = \hat{e}_p\frac{\partial\phi}{\partial x_m}\frac{\partial}{\partial x_p}\left(\frac{\partial\phi}{\partial x_m}\right),\\
&= \hat{e}_p\frac{1}{2}\frac{\partial}{\partial x_p}\left(\frac{\partial\phi}{\partial x_m}\frac{\partial\phi}{\partial x_m}\right) = \frac{1}{2}\underline{\nabla}\left[(\underline{\nabla}\phi)\cdot(\underline{\nabla}\phi)\right],\\
\Rightarrow \left[(\underline{\nabla}\phi)\cdot\underline{\nabla}\right]\underline{\nabla}\phi &= \frac{1}{2}\underline{\nabla}\left[(\underline{\nabla}\phi)\cdot(\underline{\nabla}\phi)\right].
\end{aligned} \tag{1.74}$$

Example 6. Prove that $\underline{S}:\underline{W}=0$ where $\underline{S}$ and $\underline{W}$ represent a symmetric and an antisymmetric second order tensor, respectively.

$$\begin{aligned}
\underline{S}:\underline{W} &= S_{ij}\hat{e}_i\hat{e}_j : W_{mn}\hat{e}_m\hat{e}_n = S_{ij}\hat{e}_i\cdot W_{mn}\delta_{jm}\hat{e}_n,\\
&= S_{ij}\hat{e}_i\cdot W_{jn}\hat{e}_n = S_{ij}W_{jn}\delta_{in} = S_{ij}W_{ji}.
\end{aligned} \tag{1.75}$$

The quantity on the RHS of (1.75) must remain unchanged if all i's are made j's, and all j's are made i's. Thus,

$$\underline{S}:\underline{W} = S_{ij}W_{ji} = S_{ji}W_{ij}. \tag{1.76}$$

Since $\underline{S}$ is a symmetric tensor $S_{ji}=S_{ij}$. Thus,

$$S_{ji}W_{ij} = S_{ij}W_{ij}. \tag{1.77}$$

Since $\underline{W}$ is an antisymmetric tensor $W_{ij}=-W_{ji}$. Thus,

$$S_{ij}W_{ij} = -S_{ij}W_{ji}. \tag{1.78}$$

Now equating the RHS of (1.78) directly to the right-most term of (1.75), we conclude

$$-S_{ij}W_{ji} = S_{ij}W_{ji}. \tag{1.79}$$

Equation (1.79) has both sides identical except for the negative sign. This implies,

$$S_{ij}W_{ji} = 0 \Rightarrow \underline{S}:\underline{W} = 0. \tag{1.80}$$

In the subsequent chapters of this book, we frequently refer to these useful identities (Examples 1–6) while deriving the governing equations of turbulent flows.

Description of Fluid Kinematics 2

2.1 The Continuum Description

In this book, we describe fluid motion using the *continuum* description. The continuum description does not track the motion of individual molecules but describes the motion of individual *fluid particles*. A fluid particle is assumed to be a *point mass* in the continuum description and is characterized by its velocity, acceleration (with respect to an inertial reference frame in context), density, pressure, and temperature. We refer to these quantities as the *properties* of the fluid particle. Ascertaining the values of these properties for each fluid particle in a domain of interest culminates in completely describing the motion of the fluid in that domain.

To quantify the deformation process of a fluid medium, we invoke the idea of a *fluid element*. A fluid element is a small but finite-sized chunk of mass comprised of several fluid particles. We identify a fluid element at a reference time (t), then examine how it subsequently deforms over an infinitesimal time duration Δt. Figure 2.1 shows a cuboidal fluid element identified at time t. Our working coordinate system is a frame-fixed Cartesian coordinate system $Ox_1(\hat{e}_1)x_2(\hat{e}_2)x_3(\hat{e}_3)$. At the current time t, this fluid element has a cuboidal shape with its vertices being ABCDEFGH and its edge lengths are small but finite: Δx_1, Δx_2 and Δx_3 along the directions $\hat{e}_1$, $\hat{e}_2$, and $\hat{e}_3$, respectively. At the current time instant, the coordinates of the six vertices are:

$$
\begin{aligned}
\text{Vertex A} &: (x_1, x_2, x_3),\\
\text{Vertex B} &: (x_1 + \Delta x_1, x_2, x_3),\\
\text{Vertex C} &: (x_1 + \Delta x_1, x_2 + \Delta x_2, x_3),\\
\text{Vertex D} &: (x_1, x_2 + \Delta x_2, x_3),\\
\text{Vertex E} &: (x_1, x_2, x_3 + \Delta x_3),\\
\text{Vertex F} &: (x_1 + \Delta x_1, x_2, x_3 + \Delta x_3),
\end{aligned}
$$

S. S. Sinha, *Fundamentals of Turbulence and Its Modeling*,
https://doi.org/10.1007/978-3-031-94016-3_2

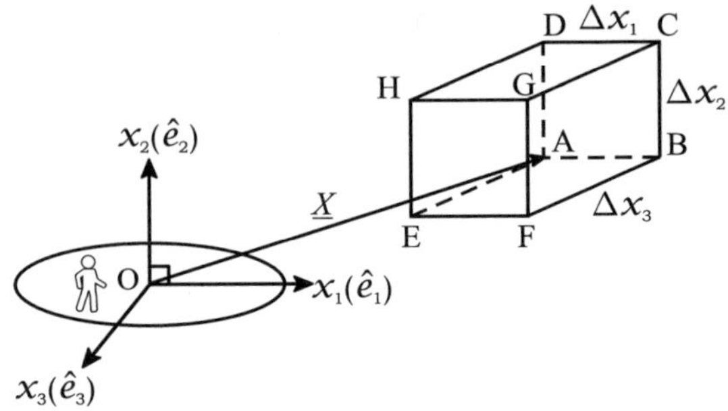

Fig. 2.1 A fluid element ABCDEFGH

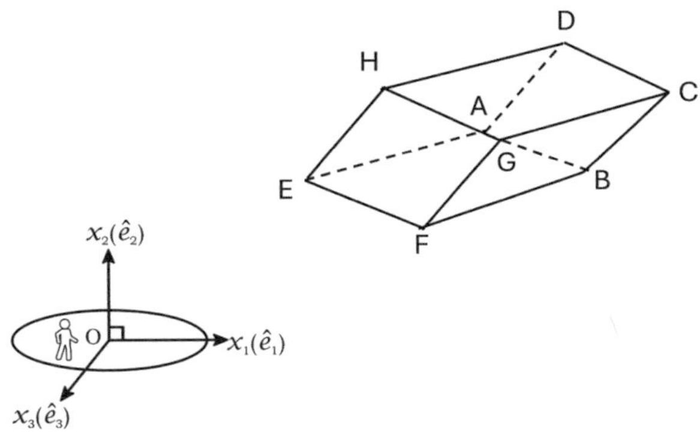

Fig. 2.2 The shape of the fluid element ABCDEFGH at $t + \Delta t$. The shape of the same fluid element was a cuboid at time t (Fig. 2.1)

$$\text{Vertex G} : (x_1 + \Delta x_1, x_2 + \Delta x_2, x_3 + \Delta x_3),$$
$$\text{Vertex H} : (x_1, x_2 + \Delta x_1, x_3 + \Delta x_3). \tag{2.1}$$

Figure 2.2 shows the same fluid element at an infinitesimal time (Δt) later at ($t + \Delta t$). In general, the six vertices have their coordinates changed compared to what they were at t. Individual vertices of the fluid element indeed translates with the local, instantaneous fluid velocity vector.

2.2 The Lagrangian Description of Fluid Continuum

The Lagrangian description chooses time (t) and the individual identity of a fluid particle to be the independent variables. Since the identity of a fluid particle is an independent variable, the Lagrangian description must devise a distinct way to name every particle in the flow domain. Even though there can be many creative ways to name individual particles, the conventional way of naming them is by the position vector of the location where a chosen fluid particle was at a fixed reference time in the past. Let us denote this reference time as t_{ref}, and let $\underline{Y}$ denote the position vector of the chosen particle at $t = t_{\text{ref}}$. The fluid particle, which was at this location at the reference time, is then referred/identified by this position vector $\underline{Y}$ at all later time instants (t, see Fig. 2.3). The velocity of this fluid particle (a dependent variable) at the current time (t) is represented symbolically as $\underline{V}^+(t, \underline{Y})$.

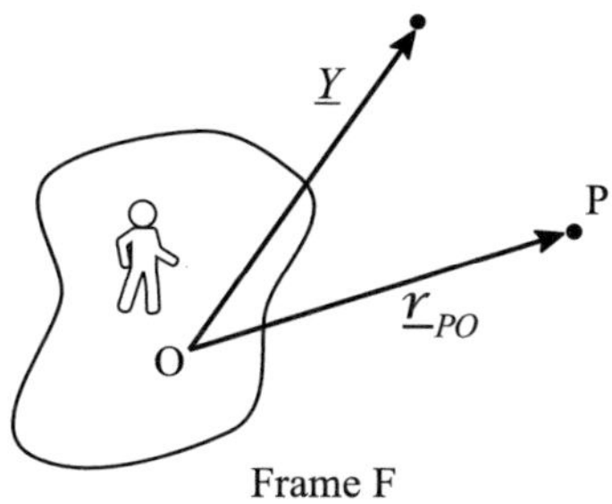

Fig. 2.3 The symbols $\underline{r}_{PO}$ and $\underline{Y}$ represent the position vectors of the same particle P at t and t_{ref}, respectively

Similarly, the density, temperature, and pressure associated with this fluid particle at the current time instant are represented as $\rho^+(t, \underline{Y})$, $T^+(t, \underline{Y})$, and $p^+(t, \underline{Y})$. It is a common practice to use the superscript + with all dependent variables while using the Lagrangian description. With $\underline{Y}$ representing the position vector of a chosen fluid particle at t_{ref}, and t representing the current time instant at which we wish to have the flow description, the symbols $\underline{V}^+(t, \underline{Y})$, $\rho^+(t, \underline{Y})$, $T^+(t, \underline{Y})$, and $p^+(t, \underline{Y})$ represent the velocity, density, temperature, and pressure *fields* of the fluid domain. The symbol $\underline{Y}$ represents a continuous spatial variable. We call $\underline{V}^+(t, \underline{Y})$ as the instantaneous (at time t) *Lagrangian velocity field*. Similarly, $\rho^+(t, \underline{Y})$, $T^+(t, \underline{Y})$, and $p^+(t, \underline{Y})$ are called the instantaneous Lagrangian fields of density, temperature and pressure respectively.

While describing fluid motion, it is of interest to inquire about the rate of change of a dependent variable of a fluid particle with time. Let us examine the partial derivative of the dependent variables of the Lagrangian description with respect to time. At time t, the partial derivative of a Lagrangian dependent variable $\phi^+(t, \underline{Y})$ with respect to time is:

$$\frac{\partial \phi^+(t, \underline{Y})}{\partial t} = \lim_{\Delta t \to 0} \frac{\phi^+(t + \Delta t, \underline{Y}) - \phi^+(t, \underline{Y})}{\Delta t}. \tag{2.2}$$

On the right-hand side, two observations of the dependent variable are being used. Since the derivative of the dependent variable is partial with respect to t, by definition, both observations have the same value of $\underline{Y}$ within the parentheses. In other words, the two observations have been made on the same fluid particle but at two different time instants: t and $t + \Delta t$. Thus, $\frac{\partial \phi^+(t, \underline{Y})}{\partial t}$ precisely represents the current rate (at time t) of change in the quantity ϕ following that fluid particle which was at location $\underline{Y}$ at the reference time (t_{ref}).

Replacing $\phi^+(t, \underline{Y})$ by the velocity vector ($\underline{V}^+(t, \underline{Y})$ in (2.2) gives us the expression for the instantaneous acceleration of the fluid particle which was at location $\underline{Y}$ at the reference time t_{ref}:

$$\underline{a}^+(t, \underline{Y}) = \frac{\partial \underline{V}^+(t, \underline{Y})}{\partial t}. \tag{2.3}$$

Similarly, the partial derivatives $\frac{\partial \rho^+(t, \underline{Y})}{\partial t}$, $\frac{\partial p^+(t, \underline{Y})}{\partial t}$, and $\frac{\partial T^+(t, \underline{Y})}{\partial t}$ represent the instantaneous rates of change in density, pressure and temperature of that fluid particle which was located at $\underline{Y}$ at t_{ref}.

In the Lagrangian description, the current location of an independently chosen fluid particle (which had its location at $\underline{Y}$ at t_{ref}) is a dependent variable. This is represented as $\underline{X}^+(t, \underline{Y})$. Indeed, this quantity is related to the time integral of the velocity vector:

$$\underline{X}^+(t, \underline{Y}) = \underline{Y} + \int_{t_{\text{ref}}}^{t} \underline{V}^+(t', \underline{Y})\, dt', \tag{2.4}$$

where $t_{\text{ref}} \leq t' \leq t$.

2.3 The Eulerian Description of Fluid Continuum

In the Eulerian description of a flow field, time (t) and an independently chosen position vector $\underline{X}$ (see Fig. 2.4) are treated as the set of independent variables. Consequently, the velocity vector $V(\underline{X}, t)$, density $\rho(\underline{X}, t)$, temperature $T(\underline{X}, t)$, and pressure $p(\underline{X}, t)$ of the fluid particle located at $\underline{X}$ at the current time t are the dependent variables. It is conventional that, unlike the Lagrangian dependent variables, the dependent variables of the Eulerian description do not have $+$ as a superscript. A further distinction is made by the order in which the independent variables appear as arguments of the dependent variables. While the Lagrangian description had the time variable appearing first and the spatial variable ($\underline{Y}$) appearing second, the Eulerian description had the spatial variable ($\underline{\text{X}}$) appearing first and the time variable appearing second. The symbol $\underline{X}$ represents a continuous spatial variable. The functions $V(\underline{X}, t)$, $\rho(\underline{X}, t)$, $T(\underline{X}, t)$, and $p(\underline{X}, t)$ are called the instantaneous Eulerian velocity, density, temperature, and pressure *fields*.

Determining the rate of change of a flow variable (such as velocity, density, temperature, and pressure) following a fluid particle is not as straightforward in the Eulerian description as it is in the Lagrangian description. In the Lagrangian approach, simply taking a partial derivative of a flow variable was sufficient (see Eq. 2.2). However, in the Eulerian description, the expression for the rate of change of $\phi(t, x_1, x_2, x_3)$ following the fluid particle located at $\underline{X}$ at time t is as follows:

$$\lim_{\Delta t \to 0} \frac{\phi(\underline{X}', t + \Delta t) - \phi(\underline{X}, t)}{\Delta t} = \frac{\partial \phi}{\partial t} + V_1 \frac{\partial \phi}{\partial x_1} + V_2 \frac{\partial \phi}{\partial x_2} + V_3 \frac{\partial \phi}{\partial x_3}, \tag{2.5}$$

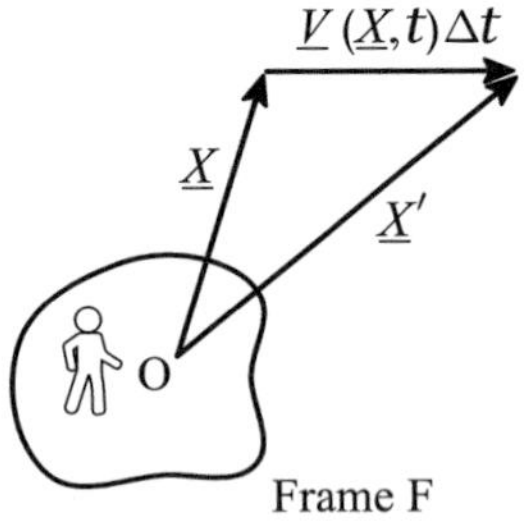

Fig. 2.4 The Eulerian description of fluid motion

where the symbol $\underline{X}^{'}$ represents the position vector at time $t + \Delta t$ of the fluid particle which was located at $\underline{X}$ at time t. The detailed derivation of (2.5) is available in [1]. The RHS of (2.5) involves not only the partial derivative of the Eulerian variable ϕ with respect to time but also the partial derivatives of the dependent variable with respect to three spatial coordinates x_1, x_2, x_3. Referring back to (1.62), (2.5) can be expressed in a form which is independent of the choice of the working coordinate system:

$$\lim_{\Delta t \to 0} \frac{\phi(\underline{X}^{'}, t + \Delta t) - \phi(\underline{X}, t)}{\Delta t} = \frac{\partial \phi(\underline{X}, t)}{\partial t} + \left(\underline{V} \cdot \underline{\nabla}\right) \phi(\underline{X}, t),$$
$$= \left(\frac{\partial}{\partial t} + \underline{V} \cdot \underline{\nabla}\right) \phi(\underline{X}, t). \tag{2.6}$$

In the study of fluid mechanics, the operator $\frac{\partial}{\partial t} + \left(\underline{V} \cdot \underline{\nabla}\right)$ is denoted by the symbol $\frac{D}{Dt}$, and is called the *material derivative* operator. It can act upon any dependent variable of the Eulerian description of a flow field. Here, the significance of the word "material" is that it represents the rate of change of the variable in context following the same fluid particle or material, which is at location $\underline{X}$ at time t.

The material derivative operator is applied on the Eulerian velocity field to arrive at the acceleration of the local fluid particle $\underline{a}(\underline{X}, t)$:

$$\underline{a}(\underline{X}, t) = \frac{D}{Dt} \underline{V}(\underline{X}, t). \tag{2.7}$$

A spatially varying Eulerian velocity field is called a non-uniform velocity field, whereas a time-dependent Eulerian velocity field is called an *unsteady* velocity field. In contrast, the Eulerian velocity field is described as *steady* if it has no time dependence. Similarly, if an Eulerian velocity field has no dependence on space, it is called a *uniform* Eulerian velocity field.

Consider an Eulerian velocity field that is expressed using a coordinate system $Ox_1(\hat{e}_1)x_2(\hat{e}_2)x_3(\hat{e}_3)$ (Fig. 1.1). If one particular scalar component of the velocity field is zero at all locations, then such a velocity field is called a *two-component* or a *2C* velocity field. Similarly, a *1-component* or a *1C* velocity field exists when two particular scalar components are identically zero at all locations. If an Eulerian velocity field depends only on two coordinates of the working coordinate system, it is called a two-dimensional or a 2D velocity field. Similarly, a one-dimensional (or 1D) velocity field is one in which the velocity field depends only on one coordinate. In general, a velocity field may depend on all three coordinates; thus, a velocity field is, in general, three-dimensional or 3D.

The Lagrangian and the Eulerian descriptions of a continuum flow field offer their own individual advantages. The Lagrangian approach appears to be more intuitive in light of our fundamental training in classical particle mechanics, wherein, indeed, the governing equations of motion are directly written for independently chosen particles (rather than spatial locations). The particle's current location of interest is calculated as a dependent variable after determining its momentum. The mathematical process of finding the rate of change of properties associated with a chosen fluid particle too is mathematically simpler in the Lagrangian description (merely, the partial derivative with time is required) compared to the Eulerian description, which involves computation of both the time and the spatial derivatives (2.6). Still, the Eulerian description has been the more popular way of describing fluid motion. This preference is attributable to the fact that for most engineering problems, our focus is indeed to measure and understand the behavior of fluid at independently chosen locations/regions rather than the behavior of specific fluid particles. Further, the expression of the forces exerted on a given fluid particle due to the interaction of the neighboring fluid particles (pressure forces and viscous forces, involves spatial gradients of the dependent variables. The Eulerian description, with the spatial location being an independent variable, simplifies the algebraic expression and the manipulation of these gradients. In the rest of this book, we describe fluid turbulence using only the Eulerian description. Further, we restrict ourselves only to those flow fields in which density and temperature are constants and the velocity vector and pressure are the only dependent field variables.

2.4 Kinematics of a Fluid Element in a 3C, 3D Velocity Field

To understand the deformations and rotations associated with a fluid element, we refer back to the fluid element of Fig. 2.1. Figure 2.2 shows its deformed state at $t + \Delta t$. In a general 3C velocity field, every constituent fluid particle of the fluid element would have its displacement vector, in general, with non-zero projections along all the three axes of the coordinate system. In general, the vertices A, B, C, and D would no longer be lying in a plane parallel to the $x_1(\hat{e}_1) - x_2(\hat{e}_2)$ plane. Similarly, the vertices, E, F, G, and H, would no more be confined to a plane parallel to the $x_1(\hat{e}_1) - x_2(\hat{e}_2)$ plane. In Fig. 2.5, we have shown projections of edges AB, AD and AE of the fluid element on three orthogonal planes at time $t + \Delta t$. These segments themselves have not been shown in the figure to avoid clutter on the figure. The projections of the segments AB and AD on the $x_1(\hat{e}_1) - x_2(\hat{e}_2)$ plane are AB' and AD', respectively. Similarly, the projections of segments AD and AE on the $x_2(\hat{e}_2) - x_3(\hat{e}_3)$ plane are AD'' and AE'', respectively. The projections of segments AE and AB on the $x_3(\hat{e}_3) - x_1(\hat{e}_1)$ plane are AE''' and AB''', respectively. Further, on these figures, we have shown various small angles that these projections make with the three axes: $x_1(\hat{e}_1)$, $x_2(\hat{e}_2)$, and $x_3(\hat{e}_3)$. The curved arrows indicate the sense in which these individual angles assume positive values. With Fig. 2.5 the reference, we can derive the various rates of geometric changes associated with the fluid element (more details available in [1]).

1. The rate of fractional change in the length AB is derived as:

$$\lim_{\Delta t \to 0} \frac{1}{\Delta t} \frac{(AB)_{t+\Delta t} - (AB)_t}{(AB)_t} = \frac{\partial V_1}{\partial x_1}. \tag{2.8}$$

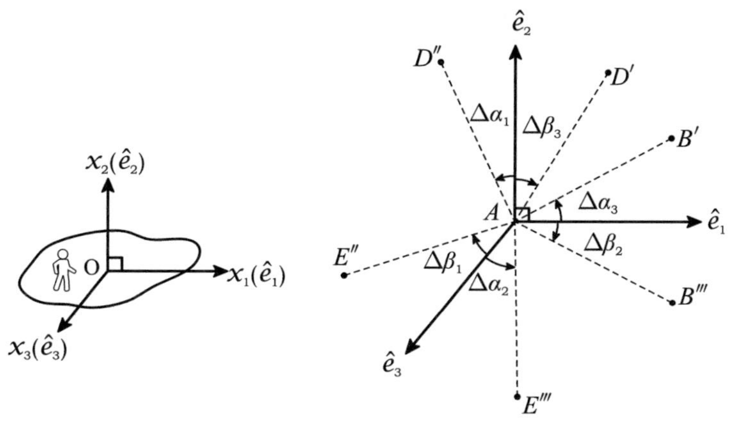

Fig. 2.5 Different projections of the edges AB, AD and AE of the fluid element ABCDEFGH on the three planes of the working coordinate system at time $t + \Delta t$. Relevant small angles and their directions are also marked on the figure. The same fluid element was a cuboid at time t (Fig. 2.1)

2. The rate of fractional change in the length AD is derived as:

$$\lim_{\Delta t \to 0} \frac{1}{\Delta t} \frac{(AD)_{t+\Delta t} - (AD)_t}{(AD)_t} = \frac{\partial V_2}{\partial x_2}. \tag{2.9}$$

3. The rate of fractional change in the length AE is derived as:

$$\lim_{\Delta t \to 0} \frac{1}{\Delta t} \frac{(AE)_{t+\Delta t} - (AE)_t}{(AE)_t} = \frac{\partial V_3}{\partial x_3}. \tag{2.10}$$

4. The rate of fractional change in the volume (dilatation rate) of the fluid element ABCDEFGH is derived as:

$$\lim_{\Delta t \to 0} \frac{1}{\Delta t} \frac{vol_{t+\Delta t} - vol_t}{vol_t} = \frac{\partial V_1}{\partial x_1} + \frac{\partial V_2}{\partial x_2} + \frac{\partial V_3}{\partial x_3}. \tag{2.11}$$

5. The rate at which the projections $AB^{'}$ and $AD^{'}$ on the $x_1(\hat{e}_1) - x_2(\hat{e}_2)$ plane tend to align with each other is derived as:

$$\lim_{\Delta t \to 0} \frac{\Delta\alpha_3 + \Delta\beta_3}{\Delta t} = \frac{\partial V_2}{\partial x_1} + \frac{\partial V_1}{\partial x_2}. \tag{2.12}$$

6. The rate at which the projections of $AD^{'}$ and $AE^{'}$ on the $x_2(\hat{e}_2) - x_3(\hat{e}_3)$ plane tend to align with each other is derived as:

$$\lim_{\Delta t \to 0} \frac{\Delta\alpha_1 + \Delta\beta_1}{\Delta t} = \frac{\partial V_3}{\partial x_2} + \frac{\partial V_2}{\partial x_3}. \tag{2.13}$$

7. The rate at which the projections $AE^{'}$ and AB' on the $x_3(\hat{e}_3) - x_1(\hat{e}_1)$ plane tend to align with each other is derived as

$$\lim_{\Delta t \to 0} \frac{\Delta\alpha_2 + \Delta\beta_2}{\Delta t} = \frac{\partial V_1}{\partial x_3} + \frac{\partial V_3}{\partial x_1}. \tag{2.14}$$

8. The component of the averaged angular velocity vector along the unit vector $\hat{e}_3$ is derived as

$$\Omega_3 = \lim_{\Delta t \to 0} \frac{\Delta\alpha_3 - \Delta\beta_3}{2\Delta t} = \frac{1}{2}\left(\frac{\partial V_2}{\partial x_1} - \frac{\partial V_1}{\partial x_2}\right). \tag{2.15}$$

9. The component of the averaged angular velocity vector along the unit vector $\hat{e}_1$ is derived as

$$\Omega_1 = \lim_{\Delta t \to 0} \frac{\Delta\alpha_1 - \Delta\beta_1}{2\Delta t} = \frac{1}{2}\left(\frac{\partial V_3}{\partial x_2} - \frac{\partial V_2}{\partial x_3}\right). \tag{2.16}$$

10. The component of the averaged angular velocity vector along the unit vector $\hat{e}_2$ is derived as

$$\Omega_2 = \lim_{\Delta t \to 0} \frac{\Delta\alpha_2 - \Delta\beta_2}{2\Delta t} = \frac{1}{2}\left(\frac{\partial V_1}{\partial x_3} - \frac{\partial V_3}{\partial x_1}\right). \tag{2.17}$$

With its three scalar components (Ω_1, Ω_2, and Ω_3) the averaged angular velocity vector itself is expressed as $\underline{\Omega} = \Omega_1\hat{e}_1 + \Omega_2\hat{e}_2 + \Omega_3\hat{e}_3$. It can be easily verified that this quantity is related to the curl of the velocity field as:

$$\underline{\Omega} = \frac{1}{2}\left(\underline{\nabla} \times \underline{V}\right). \tag{2.18}$$

In fluid mechanics, the quantity $\underline{\nabla} \times \underline{V}$ is also called *the vorticity vector*.

Further, the Cartesian components of the angular velocity vector are also related to the Cartesian components of the so-called *rotation-rate tensor* ($\underline{\text{R}}$). $\underline{R}$ is a second-order tensor and is the antisymmetric part of the velocity gradient tensor ($\underline{\nabla}\,\underline{V}$):

$$\underline{R} = \frac{(\underline{\nabla}\,\underline{V}) - (\underline{\nabla}\,\underline{V})^T}{2}. \tag{2.19}$$

and

$$\underline{R} = \Omega_3\hat{e}_1\hat{e}_2 - \Omega_2\hat{e}_1\hat{e}_3 - \Omega_3\hat{e}_2\hat{e}_1 + \Omega_1\hat{e}_2\hat{e}_3 + \Omega_2\hat{e}_3\hat{e}_1 - \Omega_1\hat{e}_3\hat{e}_2. \tag{2.20}$$

On the other hand, the symmetric part of the velocity gradient tensor is called the *strain-rate tensor* (denoted by symbol $\underline{S}$):

$$\underline{S} = \frac{(\nabla \underline{V}) + (\nabla \underline{V})^T}{2}. \tag{2.21}$$

In a Cartesian coordinate system, the strain-rate tensor is expressed as:

$$\underline{S} = S_{11}\hat{e}_1\hat{e}_1 + S_{12}\hat{e}_1\hat{e}_2 + S_{13}\hat{e}_1\hat{e}_3 + S_{21}\hat{e}_2\hat{e}_1 + S_{22}\hat{e}_2\hat{e}_2 + S_{23}\hat{e}_2\hat{e}_3 + S_{31}\hat{e}_3\hat{e}_1 + S_{32}\hat{e}_3\hat{e}_2 + S_{33}\hat{e}_3\hat{e}_3. \tag{2.22}$$

It can be verified that the various scalar components of the tensor are:

$$\begin{aligned}
S_{11} &= \frac{\partial V_1}{\partial x_1}, & S_{12} &= \frac{1}{2}\left(\frac{\partial V_2}{\partial x_1} + \frac{\partial V_1}{\partial x_2}\right), & S_{13} &= \frac{1}{2}\left(\frac{\partial V_3}{\partial x_1} + \frac{\partial V_1}{\partial x_3}\right),\\
S_{21} &= \frac{1}{2}\left(\frac{\partial V_1}{\partial x_2} + \frac{\partial V_2}{\partial x_1}\right), & S_{22} &= \frac{\partial V_2}{\partial x_2}, & S_{23} &= \frac{1}{2}\left(\frac{\partial V_3}{\partial x_2} + \frac{\partial V_2}{\partial x_3}\right),\\
S_{31} &= \frac{1}{2}\left(\frac{\partial V_1}{\partial x_3} + \frac{\partial V_3}{\partial x_1}\right), & S_{32} &= \frac{1}{2}\left(\frac{\partial V_2}{\partial x_3} + \frac{\partial V_3}{\partial x_2}\right), & S_{33} &= \frac{\partial V_3}{\partial x_3}.
\end{aligned} \tag{2.23}$$

The components S_{11}, S_{22}, and S_{33} are called the *normal* strain-rate components, whereas the other six scalar components are called the *shear* strain-rate components. Clearly, the normal strain-rate components are identical to the expressions obtained in (2.8)–(2.10). Thus, the normal components S_{11}, S_{22}, and S_{33} individually represent the rates of fractional change of the edges AB, AD, and AE, respectively. Further, we observe that the sum of the normal strain-rate components ($S_{ii} = S_{11} + S_{22} + S_{33}$) equals the dilatation rate of the fluid element ABCDEFGH (2.11). A comparison of the RHS of (2.12) with the expression of S_{12} in (2.23) shows that the shear strain-rate component S_{12} equals *half* the rate at which the projections of AB' and AD' tend to align with each other (Fig. 2.5). Similarly, the shear strain-rate component S_{23} equals *half* the rate at which the projections of AD'' and AE'' tend to align with each other. The shear strain-rate component S_{31} equals *half* the rate at which the projections of AE''' and AB''' tend to align with each other.

2.5 A Vortex in a Flow Field

A vortex (plural, *vortices*) can be described as a visually discernible coherent structure that includes a set of fluid particles moving along trajectories which curve around a common, identifiable axis. This axis may be stationary or may even be translating with respect to the reference frame in context. Further, this axis may be a straight line or a curve in the three-dimensional space of the fluid domain.

A simple and familiar example of a vortex is a *free vortex*, which exists in a potential flow field. The flow field of a free vortex is one of the basic solutions of the governing equation of an incompressible potential flow field [1]. In Fig. 2.6 we present some streamlines associated with a free vortex with its centre coinciding

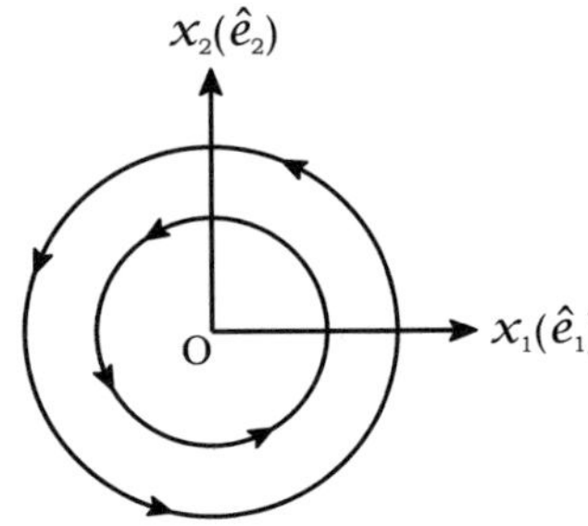

Fig. 2.6 Streamlines associated with a free vortex in a potential flow field

with the origin of the working coordinate system. In this case, the axis of this vortex is a straight line normal to the plane of the paper, and this axis remains stationary with respect to the reference frame in context. In a free vortex, all streamlines not only curve around this common axis but are also closed streamlines. Further, these streamlines are concentric circles, with the origin being their common centre. Further, in a free vortex, all the fluid particles, except the one located at the center of the vortex have zero vorticity associated with them. Even though a free vortex is a legitimate example of a vortex, we must keep in mind that, the vortices found in viscous flows (non-potential flow) may not necessarily have circular streamlines or may not even have closed streamlines. Instead, the streamlines may be spiralling around the common axis, and this common axis may not even be a straight line. Further, typically, the fluid particles that make a vortex in a viscous flow field have non-zero vorticity associated with them.

One obvious advantage of identifying vortices in a flow field is that it helps in flow visualization. Further, identifying vortices and their mutual interactions prove useful in gaining meaningful insights into the flow field leading to plausible explanations of various complex flow phenomena. Further, they help in identifying the presence of various *length-scales* and *time-scales* in the velocity field of a turbulent flow. These quantities are described in more detail Sect. 2.6. In the context of turbulent flows, a vortex is often called an *eddy* (plural, *eddies*).

2.6 Characteristic Length and Time-Scales Associated with an Eddy

Before we can describe the length scales and the time-scales associated with eddies, we adopt, for our discussion, a working definition of the *order-of-magnitude* (OM) of a number (ζ). We say that the OM of ζ is 10^M (symbolically expressed as $\mathcal{O}(\zeta) = 10^M$, or $\zeta \sim 10^M$), if

$$10^{M-0.5} \leq |\zeta| < 10^{M+0.5}, \tag{2.24}$$

where M is an integer. For example, consider the case when $\zeta = 1.5$. Since $10^{0-0.5} \leq 1.5 < 10^{0+0.5}$, following the definition (2.24), we conclude $\mathcal{O}(1.5) = 10^0 = 1$. If $\zeta = 15$, we find that $10^{1-0.5} \leq 15 < 10^{1+0.5}$, and we conclude that $\mathcal{O}(15) = 10^1 = 10$.

Closely related to the idea of the OM of a number is the characteristic value of an Eulerian field variable (ϕ). We say that ϕ_C is the characteristic value (a specific chosen number) of the variable ϕ over a domain of interest, if

$$\mathcal{O}\left[\frac{\phi}{\phi_C}\right] = 10^0 = 1, \tag{2.25}$$

over most (if not all) of the domain of interest.

Using ϕ_C, we now define a normalized version of the variable $\phi^*(x_1, x_2, x_3)$

$$\phi^* = \frac{\phi}{\phi_C}. \tag{2.26}$$

Equation (2.25) implies that if ϕ_C has been chosen aptly then

$$\mathcal{O}\left[\phi^*\right] = 1, \tag{2.27}$$

over most of the domain of interest.

The characteristic length scale of an eddy (l_{eddy}) is a number such that its order-of-magnitude is the same as the order-of-magnitude of the perceived diameter of that eddy. The characteristic time scale of an eddy (t_{eddy}) is defined as a number that has the same order-of-magnitude as that of the time it takes for a typical fluid particle located on the periphery of the eddy to spiral/revolve around the axis of the eddy (as observed by an observer who is translating with the axis of the eddy). Based on l_{eddy} and t_{eddy}, the characteristic velocity (v_{eddy}) of the eddy is defined as

$$v_{\text{eddy}} = \frac{l_{\text{eddy}}}{t_{\text{eddy}}} \tag{2.28}$$

The order-of-magnitude of v_{eddy} is the same as the OM of the velocity of a fluid particle located on the periphery of the eddy (as observed by an observer who is translating with the axis of the eddy).

2.7 An Idealized Superposition of Eddies

Let us consider a flow field wherein we have an *idealized* superposition of multiple eddies. For illustration purposes, we consider three such eddies (Fig. 2.7). These three eddies have their characteristic length scales (diameters) as $l_1 (= 2r_1)$, $l_2 (= 2r_2)$ and $l_3 (= 2r_3)$. The characteristic time scales of three eddies are t_1, t_2 and t_3, respectively. The superposition is *idealized* in the way that:

1. $C1$, $C2$ and $C3$ are particles located at the respective centres of the three eddies.
2. The centre of the second eddy (particle $C2$) is located on the periphery of the first eddy. Similarly, the center of the third eddy (particle $C2$) is located on the periphery of the second eddy.

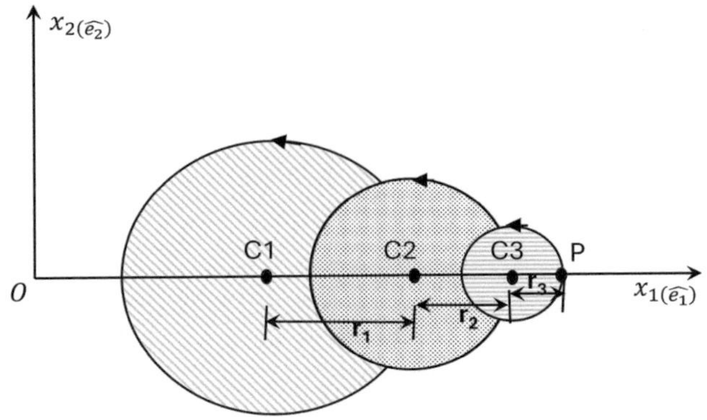

Fig. 2.7 An idealized superposition of three eddies in a flow field

3. Each eddy is assumed to behave like a rigid body. These rigid bodies (1, 2 and 3) can equivalently be treated as three different reference frames (1, 2 and 3).
4. Particle $C2$ is a common particle belonging both to body 1 and body 2. Particle $C3$ is a common particle belonging both to body 2 and body 3.
5. We assume the flow field to be planar and all three eddies have their axes along $\hat{e}_3$. Our working coordinate system is the Cartesian system $Ox_1(\hat{e}_1)x_2(\hat{e}_2)x_3(\hat{e}_3)$, which is fixed to the inertial reference frame (ground).
6. In Fig. 2.7, the fluid particle P is a part of the rigid body 3.

Here our goal is to derive the expression of the velocity of the fluid particle P, at the current instant, with respect to the ground frame using the kinematic and geometric features of the three eddies. We denote this velocity vector by $V_{P|G}$. This symbol follows the notation that $\underline{V}_{Q|\alpha}$ means the velocity of an arbitrary fluid particle Q with respect to the reference frame α [2].

With an idealized superposition of vortices as shown in Fig. 2.7, the angular velocities of the three rigid bodies can be related to the characteristic time scales of the three eddies.

$$\begin{aligned}\underline{\Omega}_{1|G} &= \frac{2\pi}{t_1}\hat{e}_3\\ \underline{\Omega}_{2|1} &= \frac{2\pi}{t_2}\hat{e}_3\\ \underline{\Omega}_{3|2} &= \frac{2\pi}{t_3}\hat{e}_3,\end{aligned} \tag{2.29}$$

where the symbol $\Omega_{\alpha|\beta}$ represents the angular velocity of frame α relative to frame β. The symbol G represents the inertial ground frame. Using the expressions in (2.29), angular velocities of the three frames relative to the ground frame are expressed as (see [2] for the full derivation of these relationships)

$$\begin{aligned}\underline{\Omega}_{3|G} &= \underline{\Omega}_{3|2} + \underline{\Omega}_{2|1} + \underline{\Omega}_{1|G}\\ \underline{\Omega}_{2|G} &= \underline{\Omega}_{2|1} + \underline{\Omega}_{1|G}\end{aligned} \tag{2.30}$$

To find $V_{P|G}$, we employ the *velocity transfer relationship* [1,2].

$$\underline{V}_{P|G} = \underline{V}_{C3|G} + \underline{\Omega}_{3|G} \times \frac{l_3}{2}\hat{e}_1. \tag{2.31}$$

Further, successively employing the velocity transfer relationship, we arrive at the following equations.

$$\underline{V}_{C3|G} = \underline{V}_{C2|G} + \underline{\Omega}_{2|G} \times \frac{l_2}{2}\hat{e}_1. \tag{2.32}$$

$$\underline{V}_{C2|G} = \underline{V}_{C1|G} + \underline{\Omega}_{1|G} \times \frac{l_1}{2}\hat{e}_1 \tag{2.33}$$

Using (2.32), (2.33) in (2.31) leads to the following expression of $\underline{V}_{P|G}$.

$$\underline{V}_{P|G} = \underline{V}_{C1|G} + \underline{\Omega}_{1|G} \times \frac{l_1}{2}\hat{e}_1 + \underline{\Omega}_{2|G} \times \frac{l_2}{2}\hat{e}_1 + \underline{\Omega}_{3|G} \times \frac{l_3}{2}\hat{e}_1. \tag{2.34}$$

Further, using the expressions of the angular velocities of various reference frames relative to the ground itself from (2.30) in (2.34) leads to the following expression of $\underline{V}_{P|G}$ at the current time instant

$$\underline{V}_{P|G} = \underline{V}_{C1|G} + \frac{\pi}{t_1}(l_1 + l_2 + l_3)\hat{e}_2 + \frac{\pi}{t_2}(l_2 + l_3)\hat{e}_2 + \frac{\pi}{t_3}l_3\hat{e}_2 \tag{2.35}$$

Equation (2.35) is the exact expression of the velocity of the particle P with respect to the ground frame at the current instant in the context of the idealized superposition of eddies considered in this section. The expression on the RHS of (2.35) clearly shows the characteristic length and time-scales of the existing eddies do influence the velocity of the particle P with respect to the ground frame. We will learn in the next chapter that every turbulent flow shows eddies of multiple time and length scales present therein. Even though all the assumptions made for our idealized superposition of eddies may not hold good in a typical turbulent flow, the kinematic expression derived in (2.35) does provide us, at least, a qualitative insight how the characteristic length and time scales of the constituent eddies in a turbulent flow field can possibly influence the velocity of fluid particles. Accordingly, a velocity field with the presence of eddies of disparate characteristic length and time scales is said to have *multiple time and length scales*.

3 Nature of Turbulent Flows

The nature of turbulent flows is often described in contrast to that of *laminar* flows. A laminar flow field may be steady or unsteady, whereas, a turbulent flow field is *always* unsteady. The velocity field in a laminar flow may be 1C, 2C or 3C, whereas the velocity field in a turbulent flow is *always* 3C. The velocity and the pressure fields in a laminar flow may be 1D, 2D or 3D, whereas in a turbulent flow field, the velocity as well as the pressure fields are *always* 3D. The vorticity field in a laminar may be zero or non-zero. A turbulent flow *always* always has a highly non-uniform and time-varying vorticity field.

A laminar flow may or may not have the presence of vortices. Even if some vortices are present in a laminar flow, there is a negligible disparity in the length and the time-scales of those resident vortices. Typically, these vortices can be described by a common characteristic length scale and a common characteristic time-scale. Furthermore, the order of the magnitude of the characteristic length scale and the characteristic time-scale are determined by the kinematic and geometric boundary conditions of the laminar flow field. For example, in the unsteady laminar flow field past a circular cylinder (of diameter D) with the velocity field in the far upstream region being uniform (V_o), vortices may periodically be shed in the wake. These vortices have a common characteristic length scale (l_{eddy}) and a common characteristic timescale (t_{eddy}), which follows

$$\mathcal{O}(l_{\text{eddy}}) = \mathcal{O}(D) \tag{3.1}$$

$$\mathcal{O}(t_{\text{eddy}}) = \mathcal{O}\left(\frac{D}{V_o}\right) \tag{3.2}$$

In contrast, a turbulent flow field is *always* perceived to have a superposition of multiple eddies, with diverse length and time-scales. We have demonstrated earlier

S. S. Sinha, *Fundamentals of Turbulence and Its Modeling*,
https://doi.org/10.1007/978-3-031-94016-3_3

in Chap. 2 (2.35) that an *idealized* superposition of multiple eddies in the flow field tends to introduce multiple time- and length-scales in the velocity field. The presence of multiple eddies in a turbulent flow is expected to introduce *a multitude of length-scales and time-scales* in the velocity field. While the characteristic length scale ($l_{\text{largest-eddy}}$) and the characteristic timescale ($t_{\text{largest-eddy}}$) of the largest eddies in a turbulent flow field are still observed to depend on the geometric and kinetic boundary conditions the flow field, (a) the ratio of the characteristic length scale of the smallest eddies to that of the largest eddies and (b) the ratio of the characteristic timescale of the smallest eddies to that of the largest eddies seem to be governed by the Reynolds number (Re_L) of the flow field. For illustration, let us again consider the flow past a circular cylinder (with D being the diameter) and the far-upstream velocity being uniform (V_0). When the flow is turbulent in the wake of the cylinder. Like in the laminar flow past the cylinder, the length scale and the timescale of the largest eddies are still determined by the boundary and initial conditions.

$$\mathcal{O}\left(l_{\text{largest-eddies}}\right) = \mathcal{O}\left(D\right) \tag{3.3}$$

and

$$\mathcal{O}\left(t_{\text{largest-eddies}}\right) = \mathcal{O}\left(\frac{D}{V_o}\right). \tag{3.4}$$

While the symbols l and t denote characteristic length and time scales, the subscript ("largest-eddies" or the "smallest eddies") denote the category of eddies in the flow field.

On the other hand, the characteristic timescale and the characteristic length scale of the smallest eddies have the following dependence:

$$\mathcal{O}\left(\frac{l_{\text{smallest-eddies}}}{l_{\text{largest-eddies}}}\right) = \mathcal{O}\left(Re_L^{-\frac{3}{4}}\right) \tag{3.5}$$

and

$$\mathcal{O}\left(\frac{t_{\text{smallest-eddies}}}{t_{\text{largest-eddies}}}\right) = \mathcal{O}\left(Re_L^{-\frac{1}{4}}\right) \tag{3.6}$$

where the definition of Re_L itself is defined as

$$Re_L = \frac{\rho V_o l_{\text{largest-eddy}}}{\mu} \tag{3.7}$$

The symbols ρ and μ represent density and the coefficient of dynamic viscosity of the fluid. The estimates of the orders-of-magnitude listed in (3.6) and (3.7) are based on the so-called *Kolmogorov hypotheses*. This will be our topic of detailed discussion later in Chap. 9.

Turbulent flows show *chaos-like* behavior. Small changes in the initial/boundary conditions associated with a turbulent flow may result into significant changes in the flow variables ($\underline{V}(\underline{X}, t)$ and $p(\underline{X}, t)$) at later times. This leads to *unpredictability* in the outcome when a turbulent flow experiment is performed next time in a laboratory.

To further explain the meaning of the word *unpredictability*, we consider an experimental set-up of a flow, with the initial and boundary conditions being specified according to a given set of measurement devices. Once the experiment starts, the flow variables may change, both in time and space. These evolving flow variables are also recorded by the same measurement devices. Consider an experiment being performed twice (or being *realized* twice) ensuring that all initial and boundary conditions are identical (as recorded by our measurement devices). Subsequently, if the recorded values of the evolved flow variables at all locations and at all subsequent time instants are found to be correspondingly identical in the two realizations of the experiment, then we say that the outcome of the experiment is *predictable* and the experiment itself is *repeatable*. Otherwise, we say the outcome of the experiment is *unpredictable* and the experiment itself is *not repeatable*. For such a flow field, the outcomes of an experiment $\underline{V}(\underline{X}, t)$ and $\underline{p}(\underline{X}, t)$ must be treated as *random* variables at every $\underline{X}$ and t. If a turbulent flow experiment starts at $t = 0$, at each location $\underline{X}$, which is within the flow domain of interest and at each time instant $t(> 0)$, we are dealing with four scalar random variables: $V_1(\underline{X}, t)$, $V_2(\underline{X}, t)$, $V_3(\underline{X}, t)$ and $\underline{p}(\underline{X}, t)$, where V_1, V_2 and V_3 are the three scalar Cartesian components of the local velocity vector ($\underline{V}(\underline{X}, t)$).

We must acknowledge, from a practical viewpoint, that the accuracy of any measurement device is always limited up to a finite number of decimal places, and some perturbations beyond those decimal places in the boundary and initial conditions are naturally always present in the environment. These facts always introduce some small differences in the initial and boundary conditions across multiple realizations of a flow experiment. Since turbulent flows show chaos-like behavior these small differences (which are not sensed by our measurement devices) across different realizations of the experiments can be amplified to the extent that the measured outcomes from these different realizations of the experiment are significantly different from each other. Thus, at local time instant from a practical viewpoint, a turbulent flow experiment is *apparently random*. Accordingly, we must treat the velocity and the pressure variables measured in a turbulent flow experiment as *random variables*. If one could build perfect measurement devices with which the initial and boundary conditions of a turbulent flow experiment can be specified exactly, and the consequent outcomes can be measured exactly, then the experiment would be repeatable, and the flow variables need not be treated as random variables. However, such perfect measurement devices do not exist.

On the other hand, a typical laminar flow does not show any chaos-like behavior. This ensures that small differences in the initial and boundary conditions (naturally existing across multiple realizations of an experiment) remain small during the subsequent evolution of the flow variables such that our measurement devices do record identical values of the flow variables in various realizations of the experiment. Thus, we do not treat the variables in a laminar flow field as random variables.

3.1 Governing Equations of Turbulent Flows

Even though several differences exist between laminar and turbulent flows, both laminar and turbulent flow fields are governed by the same set of governing equations. For a constant density flow with no heat transfer, both turbulent flows and laminar flows, when observed with respect to an inertial reference frame are governed by the Navier Stokes equation set, which is expressed (in its coordinate system independent form) as:

$$\underline{\nabla} \cdot \underline{V} = 0, \tag{3.8}$$

$$\frac{\partial \underline{V}}{\partial t} + (\underline{V} \cdot \underline{\nabla})\, \underline{V} = -\frac{1}{\rho} \underline{\nabla} p + \nu \nabla^2 \underline{V}, \tag{3.9}$$

where $\underline{V}$ and p represent the local instantaneous velocity and pressure. The symbol $\nu (= \mu/\rho)$ denotes the kinematic viscosity of the fluid and is assumed to be a constant in this book.

Equation (3.8) is a statement of mass conservation, and is commonly called the *continuity equation*. Equation (3.9) is a statement of Euler's first axiom. We refer to this equation as the *momentum equation*. The reference frame in context is an inertial reference frame. The first term on the left-hand side (LHS) of (3.9) is the unsteady term, while the second term is the advection term. On the right-hand side, the two terms represent the pressure gradient force (per unit mass) of the fluid, and the net viscous force (per unit mass). The influence of any body force has been ignored in our discussion (assumed to be of negligible importance).

If we use a Cartesian coordinate system (Fig. 1.1), which is fixed to the inertial reference frame in context, as our working coordinate system, the continuity equation is expressed as

$$\frac{\partial V_i}{\partial x_i} = 0, \tag{3.10}$$

where a repeated index implies summation (following Einstein's summation rule discussed earlier in Chap. 1). The momentum equation (3.9) is accordingly represented in the Cartesian coordinate system as

$$\frac{\partial V_i}{\partial t} + V_k \frac{\partial V_i}{\partial x_k} = -\frac{1}{\rho} \frac{\partial p}{\partial x_i} + \nu \frac{\partial^2 V_i}{\partial x_k^2} \tag{3.11}$$

Equations (3.10) and (3.11) form a set of four partial different equations (PDE) in as many unknowns (V_1, V_2, V_3 and p). Thus, the governing equation set is *mathematically closed*. While the continuity equation is a linear partial differential equation (PDE), the momentum equation (3.11) is a non-linear PDE. This non-linearity arises because of the advection term in the momentum equation.

Since the governing equation set of a turbulent flow (3.10–3.11) is mathematically closed, these partial differential equations can be solved numerically on computers using the techniques and practices of *computational fluid dynamics*. Typically, a method adopted by CFD discretizes the computational domain into small volumes (or *cells*) to convert the partial differential equations into approximate algebraic equations. Further, since all turbulent flows are always unsteady, the CFD procedure also requires time marching, for which the temporal domain is also discretized into small *timesteps*. The dimensions of these cells and the size of the timesteps must be small enough to accurately resolve the smallest length and time scales of the turbulent flow field.

Since at high Reynolds numbers, the smallest time scales and length scales of motion that exist in a turbulent flow tend to become exponentially small (3.7 and 3.6), numerically simulating a turbulent flow field requires numerous small cells and numerous small time steps to arrive at a reasonably accurate solution of the evolving flow field. Thus, the computational effort required to solve the Navier-Stokes equation for turbulent flows with adequate accuracy becomes computationally very expensive as the Reynolds number of the flow field increases. A numerical simulation adequately resolving the entire spectrum of length and time scales that exist in a turbulent flow field at the given Reynolds number is called a *direct numerical simulation (DNS)*. Even with the massive advancement of computing technology achieved in recent decades, performing direct numerical simulations is still not a viable option for many turbulent flows of engineering interest which occur at high Reynolds numbers.

The numerically computed *instantaneous* flow field, $\underline{V}(\underline{X}, t)$ and $p(\underline{X}, t)$, from a direct numerical simulation, even if available, is at best, only one possible realization of the corresponding random experiment. This is so because, like our measurement devices, the accuracy with which the initial and boundary conditions can be specified on a computer to initiate a direct numerical simulation is also finite. Due to the perturbations existing in nature, these computer-specified initial and boundary conditions may still be different from those that would exist in nature. These small differences combined with the chaos-like tendency of the turbulent flows are potent enough to drive the outcome of the actual experiment away from the DNS-computed solution. Thus, the instantaneous flow field of a DNS solution has no predictive value. However, there are still other motivations to perform accurate direct numerical simulations of turbulent flows. Some of these motivations are presented and discussed in later chapters of this book (Chaps. 5 and 8).

For many flows of practical interest, it is the Reynolds number which determines whether the flow field remains laminar or turns turbulent. For illustration, let us consider the flow of water through a pipe of diameter D. It is observed that for a pipe with a reasonably smooth internal wall, the flow of water through the pipe is laminar if the Reynolds number (Re_D) is below 2300, where Re_D is defined as

$$Re_D = UD/\nu \tag{3.12}$$

where U is expressed in terms of the volumetric flow rate (Q) through any arbitrary cross-section of the pipe,

$$U = \frac{4Q}{\pi D^2} \tag{3.13}$$

The particular value of the Reynolds number below which the pipe flow is laminar is called the *critical Reynolds number* of the flow through a pipe. Similarly, other types of flows have their own critical Reynolds number. The critical Reynolds number for the boundary layer flow over a flat plate is 5.0×10^5, approximately, where the relevant Reynolds number for the flat plate boundary layer is defined as

$$Re_x = \frac{V_o x_1}{\nu} \tag{3.14}$$

where V_o is the magnitude of the uniform velocity field in the far-upstream region, and x_1 denotes the stream-wise (along the length of the plate) distance of a location measured from the nose of the flat plate.

Even though it is observed that the laminar flow regime ceases to exist at a Reynolds number just above the respective critical value, the flow does not necessarily become turbulent rightaway. For many flows of practical interest, there exists another value of the Reynolds number (let us call it Re_T, which has a value higher than that of the critical Reynolds number) beyond which the flow becomes turbulent (exhibiting the traits described earlier in this chapter). The flow regime that exists at a Reynolds number which is larger than the critical Reynolds number of that type of flow field but is still lower than Re_T is called a *transitional* flow. Transitional flows, too, are always unsteady, but they do not exhibit the wide ranges of length and time scales that are observed in turbulent flows (3.7 and 3.6).

It is possible to calculate the critical Reynolds number for many flow fields employing the *linear stability theory*. However, an estimation of the Reynolds number beyond which a flow becomes turbulent is not amenable to any simple mathematical analysis. Indeed, even today, this topic is a subject of active research. Any further discussion on the estimation of the critical Reynolds number, estimation of Re_T or the behavior of transitional flows is deemed to be outside the scope of this book. In the rest of the text, we focus entirely on flows which have already become turbulent.

Random Variables and Their Characterization

4

In the last chapter, we discussed that despite the governing equations of a turbulent flow being known, due to the extreme dependence of the flow field on the small differences in the initial/boundary conditions, we are forced to treat turbulent flows as apparently random, and we must treat the flow variables of a turbulent flow as random variables. In this chapter, we review the essential aspects of the *probability theory* so that these flow variables can be aptly characterized. In an incompressible turbulent flow, at every time instant and at every location, there are four associated random variables: V_1, V_2, V_3 and p. While V_1, V_2 and V_3 denote the three scalar cartesian components of the local instantaneous velocity vector and p is the local instantaneous pressure.

At any chosen time instant, say t_o, the symbol $p(\underline{X}, t_o)$ denotes a set of random variables due to the continuously varying independent variable $\underline{X}$. This set of random variables is collectively referred to as the *random pressure field existing at time* t_o. Similarly, at any chosen spatial location, say $\underline{X}_o$, the symbol $p(\underline{X}_o, t)$ denotes a set of random variables due to the continuously varying independent variable t). This set of random variables is collectively called as the *random pressure process existing at the location* $\underline{X}_o$. Similarly, $V_i(\underline{X}, t_o)$ and $V_i(\underline{X}_o, t)$ denote the random field of the ith velocity component at time t_o and the random process of the ith velocity component at location $\underline{X}_o$.

4.1 An Event

We begin the review of the theory of probability in the context of a single random variable ϕ. This variable is an outcome of a random experiment, say ζ. The next time the experiment ζ is performed, we can not predict the value of ϕ with certainty. The value that the random variable ϕ actually assumes next time the experiment ζ is

S. S. Sinha, *Fundamentals of Turbulence and Its Modeling*,
https://doi.org/10.1007/978-3-031-94016-3_4

performed is called the next *realization* of the random variable ϕ. The *sample space* of the random variable ϕ is defined as the set of all possible numerical values that ϕ can take. The entire line of real numbers is taken to be the sample space of ϕ. In the context of a random experiment, an *event* is defined such that the random variable ϕ takes a value which is an element of a specific subset of the sample space. For example, an event (say Event A) is defined as

$$Event\ A : \{\phi < -5\} \tag{4.1}$$

If the next realization of ϕ (when the experiment ζ is performed next time) is such that $\phi < -5$, we say that *event A has occurred.* Similarly, we can define many other events related to our random experiment ζ:

$$Event\ B : \{3 \leq \phi < -5\} \tag{4.2}$$

We can also define events more generally: in terms of an independent variable ψ. This variable, ψ can be assigned any arbitrarily chosen value from the line of real numbers. For example,

$$Event\ C : \{\phi < \psi\} \tag{4.3}$$

4.2 Probability of an Event

The *probability* of an event is the likelihood of the occurrence of that event when the random experiment is performed next time. If B denotes an event, $P(B)$ denotes the probability of that event. By definition,

$$P(B) = \begin{cases} 1 & \text{if B is a certain event,} \\ 0 & \text{if B is an impossible event,} \\ p & \text{where, } 0<p<1 \text{ in general.} \end{cases} \tag{4.4}$$

For two events A and B, if $P(A) > P(B)$, we say that *event A is more likely to happen than B* when the random experiment is performed next time. **Examples:**

$$P(A) = 1 \text{ where Event A} : \{\phi < -\infty\} \tag{4.5}$$

$$P(B) = 0 \text{ where Event B} : \{\phi > \infty\} \tag{4.6}$$

Here event A is a certain event, whereas event B is an impossible event.

4.3 Cumulative Distribution Function of a Random Variable

The cumulative distribution function (CDF) of a random variable ϕ is defined as

$$F_{\phi}(\psi) = P\{\phi<\psi\} \tag{4.7}$$

where $P\{\phi < \psi\}$ represents the probability of the event $\phi < \psi$, and $F_{\phi}(\psi)$ is the CDF of the random variable ϕ, and ψ is an independent variable that can be assigned any arbitrarily chosen value from the real line to define a specific event. The variable ψ is called the *phase space variable* of the random variable ϕ.

It follows from the definition of CDF (4.7)

$$\begin{aligned} P\{\psi_a \leq \phi < \psi_b\} &= P\{\phi < \psi_b\} - P\{\phi < \psi_a\} \\ &= F_{\phi}(\psi = \psi_b) - F_{\phi}(\psi = \psi_a) \\ &= F_{\phi}(\psi_b) - F_{\phi}(\psi_a) \end{aligned} \tag{4.8}$$

where ψ_a and ψ_b are two numbers on the real line ($\psi_a \leq \psi_b$). Further, one can prove that the CDF has the following properties.

- If $\psi_b > \psi_a$, then it implies that $F_{\phi}(\psi_b) - F_{\phi}(\psi_a) \geq 0$. Thus, $F_{\phi}(\psi)$ is a *non-decreasing* function.
- $F_{\phi}(\psi = \infty) = 1$.
- $F_{\phi}(\psi = -\infty) = 0$.

4.4 Probability Density Function of a Random Variable

If the CDF of a random variable is differentiable, the probability density function (PDF) of the random variable ϕ is defined as

$$f_{\phi}(\psi) = \frac{dF_{\phi}(\psi)}{d\psi} = \lim_{\Delta\psi \to 0} \frac{F_{\phi}(\psi + \Delta\psi) - F_{\phi}(\psi)}{\Delta\psi} \tag{4.9}$$

where $\Delta\psi$ is a small independent change in the phase-space variable. Following the definition of PDF (4.9), $f_{\phi(\psi)}$ can be expressed in terms of the probability of an event.

$$f_{\phi}(\psi) = \lim_{\Delta\psi \to 0} \frac{F_{\phi}(\psi + \Delta\psi) - F_{\phi}(\psi)}{\Delta\psi} = \lim_{\Delta\psi \to 0} \frac{P\{\psi \leq \phi < (\psi + \Delta\psi)\}}{\Delta\psi}. \tag{4.10}$$

Further, one can prove that the PDF has the following properties.

- Since $F_{\phi}(\psi)$ is a non-decreasing function, (4.9) implies that $f_{\phi}(\psi) \geq 0$, wherever it is defined.

- $\int_{\psi_a}^{\psi_b} f_\phi(\psi)d\psi = \int_{\psi_a}^{\psi_b} dF_\phi = F_\phi(\psi_b) - F_\phi(\psi_a) = P\{\psi_a \leq \phi < \psi_b\}$, where ψ_a and ψ_b are two arbitrarily chosen values on the real number line with $\psi_a < \psi_b$.
- $\int_{-\infty}^{\infty} f_\phi(\psi)d\psi = F_\phi(\infty) - F_\phi(-\infty) = 1 - 0 = 1$.

4.5 Some Examples of Known PDFs

We present some examples of random variables with known forms of their PDFs.

Example 1. A random variable ϕ is called a *Gaussian* random variable, if its PDF has the following form,

$$f_\phi(\psi) = \frac{1}{\sigma\sqrt{2\pi}} \exp\left(-\frac{(\psi - \mu)^2}{2\sigma^2}\right) \tag{4.11}$$

where σ and μ are constants. In the special case if $\mu = 0$ and $\sigma = 1$, the random variable ϕ is called a *normal* random variable. The PDF of the Gaussian random variable is plotted in Fig. 4.1.

Example 2. A random variable ϕ is called a *uniformly* distributed random variable between two chosen numbers a and b ($a < b$), if its PDF has the following form,

$$f_\phi(\psi) = \begin{cases} \frac{1}{b-a} & \text{for } a \leq \psi \leq b \\ 0 & \text{, otherwise,} \end{cases} \tag{4.12}$$

where b and a are constants. The PDF of a uniformly distributed random variable is plotted in Fig. 4.2.

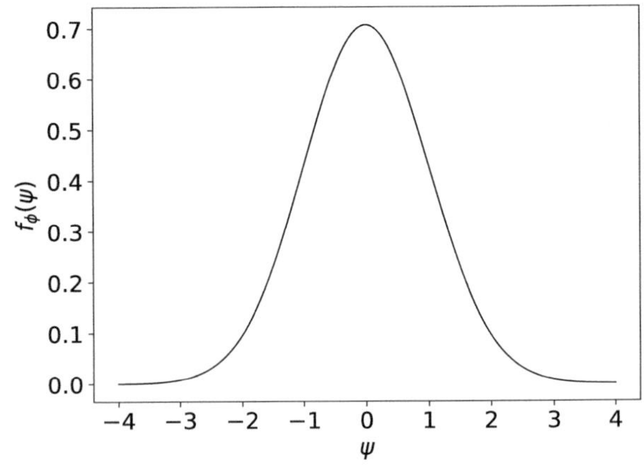

Fig. 4.1 PDF of the normal random variable

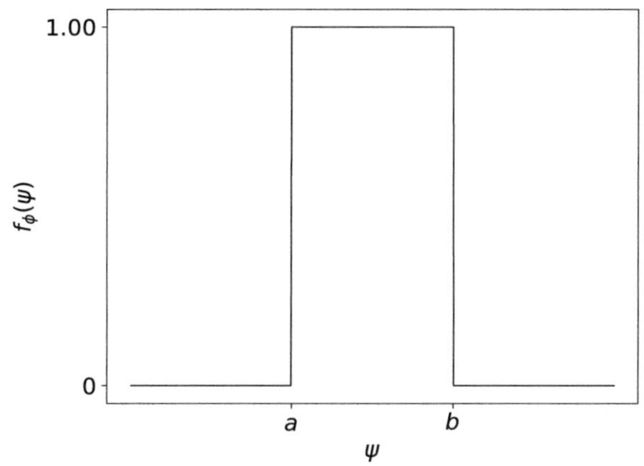

Fig. 4.2 PDF of a uniformly distributed random variable with $a = 1$ and $b = 2$

4.6 Joint CDF of Multiple Random Variables

The joint CDF of a set of N random variables $\phi_1, \phi_2 \ldots \phi_N$ is defined as

$$F_{\phi_1\phi_2\ldots\phi_N}(\psi_1, \psi_2, \ldots, \psi_N) = P\{\phi_1 < \psi_1, \phi_2 < \psi_2, \ldots, \phi_N < \psi_N\} \tag{4.13}$$

where $\psi_1, \psi_2, \ldots, \psi_N$ represent the respective phase space variables of the N random variables. The symbol $\{\phi_1 < \psi_1, \phi_2 < \psi_2, \ldots, \phi_N < \psi_N\}$ denotes an event when all the individual inequalities mentioned therein hold good simultaneously. The string appearing as the subscript of F must clearly show all the random variables jointly described by the CDF. Accordingly, in general, the CDF will be a function of the set of corresponding phase-space variables of those random variables. In particular, the joint CDF of two random variables ϕ_1 and ϕ_2 is defined as $F_{\phi_1\phi_2}(\psi_1, \psi_2)$ such that

$$F_{\phi_1\phi_2}(\psi_1, \psi_2) = P\{\phi_1 < \psi_1, \phi_2 < \psi_2\} \tag{4.14}$$

Based on the definition (4.14), it follows that

- $F_{\phi_1\phi_2}(-\infty, \psi_2) = P\{\phi_1 < -\infty \text{ and } \phi_2 < \psi_2\} = 0.$
- $F_{\phi_1\phi_2}(\infty, \psi_2) = P\{\phi_1 < \infty \text{ and } \phi_2 < \psi_2\} = P\{\phi_2 < \psi_2\} = F_{\phi_2}(\psi_2).$

In context of two random variables, $F_{\phi_2}(\psi_2)$ is also called the marginal CDF of ϕ_2. If the joint CDF is known, then the marginal CFD of either of the random variables can be determined. However, in general, we can not ascertain the joint CDF even when the individual marginal CDFs of the random variables are known.

4.7 Joint PDF of Multiple Random Variables

The joint PDF of a set of random variables is defined as

$$f_{\phi_1\phi_2...\phi_N}(\psi_1, \psi_2, ..., \psi_N) = \frac{\partial^N F_{\phi_1\phi_2...\phi_N}(\psi_1, \psi_2, ..., \psi_N)}{\partial\psi_1\partial\psi_2...\partial\psi_N} \tag{4.15}$$

if such partial derivatives of the joint CDF exist. The symbols $\psi_1, \psi_2, ..., \psi_N$ represent the respective phase space variables of the N random variables. The subscript of f must clearly show all the random variables jointly described by the PDF. Accordingly, in general, the PDF will be a function of the set of corresponding phase-space variables of those random variables. In particular, the joint PDF of two random variables ϕ_1 and ϕ_2 is defined as $f_{\phi_1\phi_2}(\psi_1, \psi_2)$ such that

$$f_{\phi_1\phi_2}(\psi_1, \psi_2) = \frac{\partial^2 F_{\phi_1\phi_2}(\psi_1, \psi_2)}{\partial\psi_1\partial\psi_2} \tag{4.16}$$

It can be proved that the joint PDF has the following properties (illustrated here for a set of two random variables ϕ_1 and ϕ_2).

1. $f_{\phi_1\phi_2}(\psi_1, \psi_2) \geq 0$, whenever it is defined.
2.
$$\int_{\psi_{1a}}^{\psi_{1b}} \int_{\psi_{2a}}^{\psi_{2b}} f_{\phi_1\phi_2}(\psi_1, \psi_2) d\psi_1 d\psi_2 = \\ \{\psi_{1a} \leq \phi_1 < \psi_{1b}, \psi_{2a} \leq \phi_1 < \psi_{2b}\} \tag{4.17}$$

   where ψ_{1a} and ψ_{1b} are two arbitrarily chosen values on the real number line with $\psi_{1a} < \psi_{1b}$. Similarly, ψ_{2a} and ψ_{2b} are two arbitrarily chosen values on the real number line with $\psi_{2a} < \psi_{2b}$.
3.
$$\int_{-\infty}^{\infty} f_{\phi_1\phi_2}(\psi_1, \psi_2) d\psi_2 = f_{\phi_1}(\psi_1). \tag{4.18}$$
4.
$$\int_{-\infty}^{\infty} f_{\phi_1\phi_2}(\psi_1, \psi_2) d\psi_1 = f_{\phi_2}(\psi_2). \tag{4.19}$$
5.
$$\int_{\infty}^{\infty} \int_{\infty}^{\infty} f_{\phi_1\phi_2}(\psi_1, \psi_2) d\psi_1 d\psi_2 = 1 \tag{4.20}$$

   where $f_{\phi_1}(\psi_1)$ and $f_{\phi_2}(\psi_2)$ are the two marginal PDFs in context.

4.8 Expected Value of a Function of Random Variables

If Q is a function of a set of random variables $\phi_1, \phi_2, ... \phi_N$, then the function Q, in general, is also a random variable. The *expected value of* Q or simply the *expectation of* Q (denoted by symbol $\langle Q \rangle$) is defined as

$$\langle Q(\phi_1, \phi_2, ...\phi_N) \rangle = \int_{-\infty}^{\infty} \int_{-\infty}^{\infty} \cdots \int_{-\infty}^{\infty} Q(\psi_1, \psi_2, \ldots, \psi_N) \, f_{\phi_1 \phi_2 \ldots \phi_N}(\psi_1, \psi_2, \ldots, \psi_N) \, d\psi_1 \, d\psi_2 \cdots d\psi_N \tag{4.21}$$

where $f_{\phi_1 \phi_2 \ldots \phi_N}(\psi_1, \psi_2, ..., \psi_N)$ is the joint PDF of the N random variables. The expected value of Q is also called the *mean* of Q.

The expected value of Q is no more a random quantity: it is a *deterministic* quantity. Its value does not depend on any individual realization of the experiment. The expected value of Q is a characterizing feature of the entire random experiment itself. The right-hand side (RHS) of Eq. (4.21) shows that $\langle Q \rangle$ is indeed the weighted integral of various possible values that Q can take in a realization of the random experiment, with the weight factor being the PDF function $f_{\phi_1 \phi_2 \ldots \phi_N}(\psi_1, \psi_2, ..., \psi_N)$. The values of random variable Q which are more probable do get more weightage in this integration process. The expectation of Q is a useful quantity for engineers, who can possibly attempt to design engineering systems based on this quantity rather than considering numerous probable values of the otherwise random quantity Q. However, the exact computation of the expected value of Q does require us to know the joint PDF of the random variables, $f_{\phi_1 \phi_2 \ldots \phi_N}(\psi_1, \psi_2, ..., \phi_N)$, in advance. In this book, we collectively refer to the expected values of various random variables and their functions as the *statistics* of that random experiment which involves these random variables and their functions.

For algebraic brevity, the expectation symbol $\langle \rangle$ is often treated like a mathematical operator that can act on various functions of one or multiple random variables. There are some properties of the expectation operator which we will frequently refer to in the later chapters of this book. Here, we list these properties along with their proofs.

4.8.1 Expectation of the Product of Q and a Non-random Quantity *a*

$$\langle aQ \rangle = a \langle Q \rangle \tag{4.22}$$

where Q denotes a function of N random variables (unless specified otherwise), and a is a non-random quantity which may still be a function of time and space.

Proof:

$$
\begin{aligned}
&\langle aQ\rangle \\
&= \int_{-\infty}^{\infty}\int_{-\infty}^{\infty}...\int_{-\infty}^{\infty} aQ(\psi_1, \psi_2, ..., \psi_N) f_{\phi_1\phi_2...\phi_N}(\psi_1, \psi_2, ...d\phi_N)d\psi_1 d\psi_2...\psi_N \\
&= a\int_{-\infty}^{\infty}\int_{-\infty}^{\infty}...\int_{-\infty}^{\infty} Q(\psi_1, \psi_2, ..., \psi_N) f_{\phi_1\phi_2...\phi_N}(\psi_1, \psi_2, ...d\phi_N)d\psi_1 d\psi_2...\psi_N \\
&= a\langle Q\rangle
\end{aligned}
\tag{4.23}
$$

where a being a non-random quantity is not dependent on the phase-space variables $\psi_1, \psi_2, ..., \psi_N$, and it has thus been taken out of the integration process in (4.23).

4.8.2 Expectation of the Expectation of Q

$$
\langle\langle Q\rangle\rangle = \langle Q\rangle. \tag{4.24}
$$

Proof:

$$
\langle\langle Q\rangle\rangle = \int_{-\infty}^{\infty}\int_{-\infty}^{\infty}\cdots\int_{-\infty}^{\infty} \langle Q\rangle f_{\phi_1\phi_2...\phi_N}(\psi_1, \psi_2, \ldots, \psi_N) \\
d\psi_1\, d\psi_2 \cdots d\psi_N.
$$

Since $\langle Q\rangle$ is a non-random quantity, it can be pulled out of the integration process, leading to

$$
\begin{aligned}
\langle\langle Q(\phi_1, \phi_2, \ldots, \phi_N)\rangle\rangle &= \langle Q\rangle \int_{-\infty}^{\infty}\int_{-\infty}^{\infty}\cdots\int_{-\infty}^{\infty} f_{\phi_1\phi_2...\phi_N}(\psi_1, \psi_2, \ldots, \psi_N) \\
&\quad \times d\psi_1\, d\psi_2 \cdots d\psi_N \\
&= \langle Q\rangle
\end{aligned}
$$

where (4.20) has been employed.

4.8.3 Expected Value of a Random Variable ϕ_1

When $Q = \phi_1$, we call the corresponding expectation (4.21) as the *expected value of the random variable* ϕ_1.

$$
\langle\phi_1\rangle = \int_{-\infty}^{\infty} \phi_1 f_{\phi_1}(\psi_1) d\phi_1, \tag{4.25}
$$

where $\langle\phi_1\rangle$ is the expected value of ϕ_1.

Example 1:
The expected value of a Gaussian random variable (PDF as described in (4.11)) is determined as

$$\begin{aligned}
\langle \phi_1 \rangle &= \int_{-\infty}^{\infty} \psi_1 f_{\phi_1}(\psi_1)\, d\psi_1 \\
&= \int_{-\infty}^{\infty} \psi_1 \frac{1}{\sigma\sqrt{2\pi}} \exp\left(-\frac{(\psi_1-\mu)^2}{2\sigma^2}\right) d\psi_1 \\
&= \frac{1}{\sigma\sqrt{2\pi}} \int_{-\infty}^{\infty} \psi_1 \exp\left(-\frac{(\psi_1-\mu)^2}{2\sigma^2}\right) d\psi_1 \\
&= \mu
\end{aligned} \tag{4.26}$$

Example 2:
The expected value of a uniformly distributed random variable ($a < b$, PDF as described in (4.12)) is determined as

$$\begin{aligned}
\langle \phi_1 \rangle &= \int_{-\infty}^{\infty} \psi_1 f_{\phi_1}(\psi_1)\, d\psi_1 \\
&= \int_{-\infty}^{a} \psi_1 f_{\phi_1}(\psi_1)\, d\psi_1 + \int_{a}^{b} \psi_1 f_{\phi_1}(\psi_1)\, d\psi_1 + \int_{b}^{\infty} \psi_1 f_{\phi_1}(\psi_1)\, d\psi_1 \\
&= 0 + \int_a^b \psi_1 \frac{1}{b-a}\, d\psi_1 + 0 \\
&= \frac{1}{b-a} \int_a^b \psi_1\, d\psi_1 \\
&= \frac{1}{b-a} \left[\frac{\psi_1^2}{2}\right]_a^b \\
&= \frac{1}{b-a} \left(\frac{b^2}{2} - \frac{a^2}{2}\right) \\
&= \frac{b+a}{2}
\end{aligned} \tag{4.27}$$

4.8.4 Expected Value of a Sum of Random Variables

$$\langle \phi_1 + \phi_2 + \ldots + \phi_N \rangle = \langle \phi_1 \rangle + \langle \phi_2 \rangle + \ldots \langle \phi_N \rangle. \tag{4.28}$$

Proof:
We provide the proof for the sum of two random variables, ϕ_1 and ϕ_2. This procedure can be easily extended to the sum of any number of random variables. We first define $Q(\phi_1, \phi_2) = \phi_1 + \phi_2$. Thus,

$$\begin{aligned}
\langle \phi_1 + \phi_2 \rangle &= \langle Q \rangle \\
&= \int_{-\infty}^{\infty} \int_{-\infty}^{\infty} Q f_{\phi_1\phi_2}(\psi_1, \psi_2)\, d\psi_1\, d\psi_2 \\
&= \int_{-\infty}^{\infty} \int_{-\infty}^{\infty} (\psi_1 + \psi_2) f_{\phi_1\phi_2}(\psi_1, \psi_2)\, d\psi_1\, d\psi_2 \\
&= \int_{-\infty}^{\infty} \int_{-\infty}^{\infty} \psi_1 f_{\phi_1\phi_2}(\psi_1, \psi_2)\, d\psi_1\, d\psi_2 \\
&\quad + \int_{-\infty}^{\infty} \int_{-\infty}^{\infty} \psi_2 f_{\phi_1\phi_2}(\psi_1, \psi_2)\, d\psi_1\, d\psi_2 \\
&= \int_{-\infty}^{\infty} \psi_1 \left[\int_{-\infty}^{\infty} f_{\phi_1\phi_2}(\psi_1, \psi_2)\, d\psi_2 \right] d\psi_1 \\
&\quad + \int_{-\infty}^{\infty} \psi_2 \left[\int_{-\infty}^{\infty} f_{\phi_1\phi_2}(\psi_1, \psi_2)\, d\psi_1 \right] d\psi_2
\end{aligned} \tag{4.29}$$

Using (4.18) and (4.19), (4.29) simplifies to

$$\begin{aligned}
\langle \phi_1 + \phi_2 \rangle &= \int_{-\infty}^{\infty} \psi_1 f_{\phi_1}(\psi_1)\, d\psi_1 + \int_{-\infty}^{\infty} \psi_2 f_{\phi_2}(\psi_2)\, d\psi_2 \\
\langle \phi_1 + \phi_2 \rangle &= \langle \phi_1 \rangle + \langle \phi_2 \rangle
\end{aligned} \tag{4.30}$$

4.8.5 Fluctuation in a Random Variable ϕ_1

The difference between the random variable ϕ_1 and its expected value is called the *fluctuation* in ϕ_1. It is denoted by the symbol ϕ_1'.

$$\phi_1' = \phi_1 - \langle \phi_1 \rangle \tag{4.31}$$

Like ϕ_1, ϕ_1', too, is a random variable. The expected value of the fluctuation is always zero.

$$\langle \phi_1' \rangle = 0. \tag{4.32}$$

Proof:

$$\langle \phi_1' \rangle = \langle \phi_1 - \langle \phi_1 \rangle \rangle \tag{4.33}$$

Using (4.30) in the RHS of (4.33) we arrive at

$$\langle \phi_1' \rangle = \langle \phi_1 \rangle - \langle \langle \phi_1 \rangle \rangle \tag{4.34}$$

Further, using (4.24), (4.34) simplifies to

$$\langle \phi_1' \rangle = \langle \phi_1 \rangle - \langle \phi_1 \rangle = 0. \tag{4.35}$$

4.8.6 Expectation of the Product of Two Random Variables

$$\langle \phi_1 \phi_2 \rangle = \langle \phi_1 \rangle \langle \phi_2 \rangle + \langle \phi_1^{'} \phi_2^{'} \rangle \tag{4.36}$$

Proof:
First, in accordance with (4.31), we substitute ϕ_1 and ϕ_2 in terms of the corresponding expected values and fluctuations.

$$\begin{aligned} \langle \phi_1 \phi_2 \rangle &= \langle (\langle \phi_1 \rangle + \phi_1^{'})(\langle \phi_2 \rangle + \phi_2^{'}) \rangle \\ &= \langle \langle \phi_1 \rangle \langle \phi_2 \rangle + \langle \phi_1 \rangle \phi_2^{'} + \phi_1^{'} \langle \phi_2 \rangle + \phi_1^{'} \phi_2^{'} \rangle \end{aligned} \tag{4.37}$$

Using (4.28) in (4.36) leads to

$$\begin{aligned} \langle \phi_1 \phi_2 \rangle = {} & \langle \langle \phi_1 \rangle \langle \phi_2 \rangle \rangle + \langle \langle \phi_1 \rangle \phi_2^{'} \rangle \\ & + \langle \phi_1^{'} \langle \phi_2 \rangle \rangle + \langle \phi_1^{'} \phi_2^{'} \rangle \end{aligned} \tag{4.38}$$

Further, using (4.23) and (4.35) simplifies (4.38) to

$$\begin{aligned} \langle \phi_1 \phi_2 \rangle = {} & \langle \phi_1 \rangle \langle \phi_2 \rangle + \langle \phi_1 \rangle \langle \phi_2^{'} \rangle + \langle \phi_1^{'} \rangle \langle \phi_2 \rangle \\ & + \langle \phi_1^{'} \phi_2^{'} \rangle \end{aligned} \tag{4.39}$$

Finally using (4.32) in (4.39) leads to

$$\langle \phi_1 \phi_2 \rangle = \langle \phi_1 \rangle \langle \phi_2 \rangle + \langle \phi_1^{'} \phi_2^{'} \rangle$$

where $\langle \phi_1^{'} \phi_2^{'} \rangle$ is the expectation of the product of the fluctuations of the two random variables ϕ_1 and ϕ_2. In general, $\langle \phi_1^{'} \phi_2^{'} \rangle \neq 0$. If required, the expression of the expectation of the product of a higher number of random variables can similarly be derived.

4.8.7 Moments of a Random Variable

The nth-*order raw moment* of the random variable ϕ_1 is defined as

$$\langle \phi_1^n \rangle = \int_{-\infty}^{\infty} f_{\phi_1} \psi_1^n d\psi_1. \tag{4.40}$$

The nth-*order central moment* of the random variable ϕ_1 is defined as

$$\langle (\phi_1 - \langle \phi_1 \rangle)^n \rangle = \int_{-\infty}^{\infty} (\psi_1 - \langle \phi_1 \rangle)^n f_{\phi_1} d\psi_1. \tag{4.41}$$

If $n = 2$, then the corresponding central moment is called the *variance* of ϕ_1, and is denoted by $var(\phi_1)$

$$var(\phi_1) = \langle(\phi_1 - \langle\phi_1\rangle)^2\rangle = \langle\phi_1^{'}\phi_1^{'}\rangle = \langle\phi_1^{'2}\rangle \tag{4.42}$$

The square-root of the variance of ϕ_1 is called the *standard deviation* of the random variable.

Example 1:
The variance of a Gaussian random variable (PDF as described in (4.11)) is determined as

$$\begin{aligned}
\langle\phi_1^{'}\phi_1^{'}\rangle &= \int_{-\infty}^{\infty} (\psi_1 - \langle\phi_1\rangle)^2 f_{\phi_1}(\psi_1)\, d\psi_1 \\
&= \int_{-\infty}^{\infty} (\psi_1 - \mu)^2 f_{\phi_1}(\psi_1)\, d\psi_1 \\
&= \int_{-\infty}^{\infty} (\psi_1 - \mu)^2 \frac{1}{\sigma\sqrt{2\pi}} \exp\left(-\frac{(\psi_1 - \mu)^2}{2\sigma^2}\right) d\psi_1 \\
&= \frac{1}{\sigma\sqrt{2\pi}} \int_{-\infty}^{\infty} (\psi_1 - \mu)^2 \exp\left(-\frac{(\psi_1 - \mu)^2}{2\sigma^2}\right) d\psi_1 \\
&= \sigma^2.
\end{aligned} \tag{4.43}$$

Example 2:
The variance of a uniformly distributed random variable between those chosen numbers a and b ($a < b$ with the PDF as described in (4.12)) is determined as

$$\begin{aligned}
\langle\phi_1^{'}\phi_1^{'}\rangle &= \int_{-\infty}^{\infty} (\psi_1 - \langle\phi_1\rangle)^2 f_{\phi_1}(\psi_1)\, d\psi_1 \\
&= \int_{-\infty}^{a} \left(\psi_1 - \frac{a+b}{2}\right)^2 f_{\phi_1}(\psi_1)\, d\psi_1 \\
&+ \int_{a}^{b} \left(\psi_1 - \frac{a+b}{2}\right)^2 f_{\phi_1}(\psi_1)\, d\psi_1 \\
&+ \int_{b}^{\infty} \left(\psi_1 - \frac{a+b}{2}\right)^2 f_{\phi_1}(\psi_1)\, d\psi_1 \\
&= 0 + \int_{a}^{b} \left(\psi_1 - \frac{a+b}{2}\right)^2 \frac{1}{b-a}\, d\psi_1 + 0 \\
&= \frac{(b-a)^2}{12}
\end{aligned} \tag{4.44}$$

For a set of two random variables, ϕ_1 and ϕ_2, we define *covariance* of ϕ_1 and ϕ_2 (denoted by the symbol $cov(\phi_1, \phi_2)$) as

$$cov(\phi_1, \phi_2) = \langle(\psi_1 - \langle\phi_1\rangle)(\psi_2 - \langle\phi_2\rangle)\rangle = \langle\phi_1^{'}\phi_2^{'}\rangle$$

$$= \int_{-\infty}^{\infty}\int_{-\infty}^{\infty}(\psi_1 - \langle\phi_1\rangle)(\psi_2 - \langle\phi_2\rangle) f_{\phi_1\phi_2}(\psi_1, \psi_2) d\psi_1 d\psi_2 \tag{4.45}$$

The *correlation coefficient* of ϕ_1 and ϕ_2 is defined as

$$\rho_{\phi_1\phi_2} = \frac{\langle\phi_1'\phi_2'\rangle}{\sqrt{\langle\phi_1'\phi_1'\rangle\langle\phi_2'\phi_2'\rangle}} \tag{4.46}$$

4.9 Expectation of Derivatives of Random Processes and Random Fields

Let $\phi(\underline{X}, t)$ denote a time-varying random field (and a random process). In the Eulerian description, the spatial position vector $(\underline{X})$ and time (t) are independent variables (Sect. 2.3).

The symbol $\langle\phi(\underline{X}, t)\rangle$ denotes the expected value of that specific random variable which is associated with the random field at an arbitrarily chosen location $\underline{X}$ and at an arbitrarily chosen time instant t. In general, $\langle\phi(\underline{X}, t)\rangle$, which is a non-random quantity, can still vary with time and space. The following relationships exist in the context of expectations of the partial derivatives of the flow variables.

4.9.1 Partial Derivative with Respect to Time

$$\left\langle\frac{\partial}{\partial t}\phi\left(\underline{X}, t\right)\right\rangle = \frac{\partial}{\partial t}\left\langle\phi\left(\underline{X}, t\right)\right\rangle \tag{4.47}$$

Proof:

$$\left\langle\frac{\partial}{\partial t}\phi\left(\underline{X}, t\right)\right\rangle = \left\langle\lim_{\Delta t\to o}\frac{\phi(\underline{X}, t + \Delta t) - \phi(\underline{X}, t)}{\Delta t}\right\rangle \tag{4.48}$$

where Δt is an arbitrary increment in t. The expectation operator involves integration over the phase space variables, and thus time the position vectors are treated as constants during this integration process. Thus using (4.28), (4.48) is expressed as

$$\begin{aligned}\left\langle\frac{\partial}{\partial t}\phi\left(\underline{X}, t\right)\right\rangle &= \lim_{\Delta t\to o}\frac{\langle\phi(\underline{X}, t + \Delta t)\rangle - \langle\phi(\underline{X}, t)\rangle}{\Delta t}\\ &= \frac{\partial}{\partial t}\langle\phi(\underline{X}, t)\rangle\end{aligned} \tag{4.49}$$

4.9.2 Partial Derivative with Respect to Spatial Coordinates

$$\left\langle \frac{\partial}{\partial x_i} \phi\left(x_1, x_2, x_3, t\right) \right\rangle = \frac{\partial}{\partial x_i} \left\langle \phi\left(x_1, x_2, x_3, t\right) \right\rangle \tag{4.50}$$

where x_i is the ith spatial coordinate of the position vector $\underline{X}$ in a frame-fixed Cartesian coordinate system $Ox_1(\hat{e}_1)x_2(\hat{e}_2)x_3(\hat{e}_3)$ (Fig. 1.1). Using this coordinate system, we express

$$\underline{X} = x_1\hat{e}_1 + x_2\hat{e}_2 + x_3\hat{e}_3 \text{ and } \phi(\underline{X}, t) = \phi(x_1, x_2, x_3, t). \tag{4.51}$$

Here we provide the proof of (4.50) for the case $x_i = x_1$.

Proof:

$$\left\langle \frac{\partial}{\partial x_1} \phi(x_1, x_2, x_3, t) \right\rangle = \left\langle \lim_{\Delta x_1 \to 0} \frac{\phi(x_1 + \Delta x_1, x_2, x_3, t) - \phi(x_1, x_2, x_3, t)}{\Delta x_1} \right\rangle \tag{4.52}$$

where Δx_1 is an arbitrary increment in x_1. Using (4.28), (4.52) is expressed as

$$\begin{aligned} \left\langle \frac{\partial}{\partial x_1} \phi\left(x_1, x_2, x_3, t\right) \right\rangle &= \lim_{\Delta x_1 \to o} \frac{\langle \phi(x_1 + \Delta x_1, x_2, x_3, t)\rangle - \langle \phi(x_1, x_2, x_3, t)\rangle}{\Delta x_1} \\ &= \frac{\partial}{\partial x_1} \langle \phi(x_1, x_2, x_3, t)\rangle \end{aligned} \tag{4.53}$$

4.10 Categorization of Turbulent Flows Based on Their Statistics

A turbulent flow is called *statistically stationary* if all statistics at every location in the flow are independent of time. These statistics of the flow may still vary with location. A turbulent flow is called *statistically homogeneous*, if all statistics at every time instant are identical at all locations in the flow domain. These statistics of the flow may still vary with time. A turbulent flow is called *statistically homogeneous along a specific line*, if all statistics at every time instant are identical at all locations on that line. These statistics of the flow may still vary with time. A turbulent flow is called *statistically homogeneous on a specific area*, if all statistics at every time instant are identical at all locations on that area. These statistics of the flow may still vary with time.

4.11 Expectation and Averaging

The expectation of a random variable or that of a function of a random variable(s) is defined as (4.21). However, determination of the expected value of a quantity following this definition would require us to be aware of the PDF (or the joint PDF) function appearing in the integral on the RHS of (4.21). In our study of turbulent flows, however, such PDFs or the joint PDFs of the flow variables are rarely known in advance. Thus, we cannot exactly calculate the expected values of the random variables by performing the integral on the RHS of (4.21). Therefore, we must explore alternate ways to find, at least approximately, the expected values of random variables of interest. There are certain kinds of averaging processes with which one can estimate these expected values. However, care must be taken in selecting the appropriate averaging procedure. Some of these averaging procedures can be employed for estimating various expected values only for certain types of turbulent flows.

4.11.1 Ensemble Averaging

Consider the random variable at an arbitrarily chosen location ($\underline{X}_o$) and at an arbitrarily chosen time instant (t_o).That random variable is represented as $\phi(\underline{X} = \underline{X}_o, t = t_o)$ or simply as $\phi(\underline{X}_o, t_o)$, where $\phi(\underline{X}, t)$ is an Eulerian variable of interest (such as a scalar component of the velocity vector or pressure). The ensemble average of the random variable $\phi(\underline{X}_o, t_o)$ is defined as

$$\langle \phi(\underline{X}_o, t_o) \rangle_N = \frac{1}{N} \sum_{i=1}^{N} {}^i\phi(\underline{X}_o, t_o) \tag{4.54}$$

where ${}^i\phi(\underline{X}_o, t_o)$ denotes the realized value of the random variable $\phi(\underline{X}_o, t_o)$ in the ith realization of the experiment, and N denotes the total number of times the turbulent flow experiment has been performed. The relationship between the ensemble average and the expectation of the random variable $\phi(\underline{X}_o, t_o)$ is

$$\lim_{N\to\infty} \langle \phi(\underline{X}_o, t_o) \rangle_N \to \langle \phi(\underline{X}_o, t_o) \rangle \tag{4.55}$$

The relationship (4.55) holds good without any restrictions to all types of turbulent flows.

4.11.2 Time-Averaging

We construct a time-averaged quantity (denoted by the symbol $\langle \phi(\underline{X}_o) \rangle_T$) associated with the random process $\phi(\underline{X} = \underline{X}_o, t)$ where t is the independent time variable and $\underline{X}_o$ represents an arbitrarily chosen location in the flow field.

$$\left\langle \phi(\underline{X}_o) \right\rangle_T = \frac{1}{T} \int_{T_1}^{T_2} \phi(\underline{X}_o, t) dt \tag{4.56}$$

where $T = T_2 - T_1$. The symbols T_1 and T_2 are the time instants at which the integration process starts and ends, respectively. *If the turbulent flow field is known to be statistically stationary* over the time period T, then $\left\langle \phi(\underline{X}_o) \right\rangle_T$ can be used to estimate the expected value of $\phi(\underline{X}_o, t)$ at any arbitrarily chosen location $\underline{X}_o$ and at any time $t \in [T_1, T_2]$.

$$\lim_{T\to\infty} \left\langle \phi(\underline{X}_o) \right\rangle_T \to \left\langle \phi(\underline{X}_o, t) \right\rangle \tag{4.57}$$

4.11.3 Line Averaging

Let us identify a line (denoted by $\mathcal{L}$ and having its length L) existing in the domain of a turbulent flow. We construct a line-averaged quantity (denoted by the symbol $\langle \phi(t_o) \rangle_{\mathcal{L}}$) at an arbitrarily chosen time instant, t_o:

$$\langle \phi(t_o) \rangle_{\mathcal{L}} = \frac{1}{L} \int_{\mathcal{L}} \phi(\underline{X}, t_o) dL \tag{4.58}$$

where dL represents an infinitesimal segment on the line $\mathcal{L}$. *If the turbulent flow field is known to be statistically homogeneous along* $\mathcal{L}$, then $\langle \phi(t_o) \rangle_L$ can be used to estimate the expected value of $\phi(\underline{X}, t_o)$ at any location on $\mathcal{L}$ and at any arbitrarily chosen time t_o.

$$\lim_{L\to\infty} \langle \phi(t_o) \rangle_{\mathcal{L}} \to \left\langle \phi(\underline{X}, t_o) \right\rangle \tag{4.59}$$

4.11.4 Area Averaging

Let us identify an area (denoted by $\mathcal{A}$ and having its magnitude A) existing in the domain of a turbulent flow. We construct an area-averaged quantity (denoted by the symbol $\langle \phi(t_o) \rangle_{\mathcal{A}}$) at an arbitrarily chosen time instant, t_o:

$$\langle \phi(t_o) \rangle_{\mathcal{A}} = \frac{1}{A} \int_{\mathcal{A}} \phi(\underline{X}, t_o) dA \tag{4.60}$$

where dA represents an infinitesimal part of the area $\mathcal{A}$. *If the turbulent flow field is known to be statistically homogeneous on* $\mathcal{A}$, then $\langle \phi(t_o) \rangle_{\mathcal{A}}$ can be used to estimate the expected value of $\phi(\underline{X}, t_o)$ at any location on $\mathcal{A}$ and at any arbitrarily chosen time t_o.

$$\lim_{\mathcal{A}\to\infty} \langle \phi(t_o) \rangle_{\mathcal{A}} \to \left\langle \phi(\underline{X}, t_o) \right\rangle \tag{4.61}$$

4.11.5 Volume Averaging

To define volume averaging, let us identify a volume (denoted by $\mathcal{V}$ and having its volume V) existing in the domain of a turbulent flow. We construct a volume-averaged quantity (denoted by the symbol $\langle\phi(t_o)\rangle_{\mathcal{V}}$) at an arbitrarily chosen time instant, t_o:

$$\langle\phi(t_o)\rangle_{\mathcal{V}} = \frac{1}{V}\int_{\mathcal{V}} \phi(\underline{X}, t_o)dV \tag{4.62}$$

where dV represents an infinitesimal part of the volume $\mathcal{V}$. *If the turbulent flow field is known to be statistically homogeneous in* $\mathcal{V}$, then $\langle\phi(t_o)\rangle_{\mathcal{V}}$ can be used to estimate the expected value of $\phi(\underline{X}, t_o)$ at any location in the region $\mathcal{V}$ and at any arbitrarily chosen time t_o.

$$\lim_{\mathcal{V}\to\infty} \langle\phi(t_o)\rangle_{\mathcal{V}} \to \big\langle\phi(\underline{X}, t_o)\big\rangle \text{ on volume } \mathcal{V}. \tag{4.63}$$

Even though the ensemble-averaging is the most versatile type of averaging (it can theoretically be applied to any type of turbulent flow field), it is often not employed to estimate the expected values of flow variables or their functions. Ensemble averaging would require repeating the turbulent flow experiment a very large number of times (4.55), which is often quite impractical to implement. On the other hand, if the flow field of interest is known to be statistically homogeneous/stationary, then we can perform that experiment only once and record our measurements. Subsequently, we can use the appropriate type of averaging (line/area/volume/time-averaging) to approximately find the expected values of the quantities of interest. Indeed, this is a common practice followed by experimentalists. The expected values of various flow variables and functions can also be estimated by performing the appropriate type of averaging (line/area/volume/time) of the database available from direct numerical simulations, provided the flow field is known to be statistically homogeneous along a line/area/volume or already known to be statistically stationary.

Governing Equations of the Mean Flow Field

5

In this chapter, we derive the governing equations of the mean pressure and velocity fields. The governing equations of $\langle \underline{V}(\underline{X}, t) \rangle$ and $\langle p(\underline{X}, t) \rangle$ can possibly provide us a direct way to compute these flow statistics without relying on any averaging procedure (discussed in Sect. 4.11). Further, these equations can help us gain deeper insights into what influences the evolution of the mean flow field in a turbulent flow.

To derive the governing equations $\langle \underline{V}(\underline{X}, t) \rangle$ and $\langle p(\underline{X}, t) \rangle$ we begin with the governing equations of $\underline{V}(\underline{X}, t)$ and $p(\underline{X}, t)$ (3.8 and 3.9). To clearly distinguish them from the equations that we plan to derive for the mean variables, we refer to (3.10) and (3.11) as the *instantaneous Navier-tokes equation set*, and the variables $\underline{V}(\underline{X}, t)$ and $p(\underline{X}, t)$ appearing therein are called the *instantaneous velocity vector* and *instantaneous pressure*, respectively. As discussed in the last chapter, the instantaneous flow variables are treated as random variables.

5.1 The Mean Continuity Equation

We first subject the continuity equation (3.10) to the mean operator

$$\left\langle \frac{\partial V_i}{\partial x_i} \right\rangle = \langle 0 \rangle \text{ or,}$$

$$\left\langle \frac{\partial V_1}{\partial x_1} + \frac{\partial V_2}{\partial x_2} + \frac{\partial V_3}{\partial x_3} \right\rangle = 0. \tag{5.1}$$

S. S. Sinha, *Fundamentals of Turbulence and Its Modeling*,
https://doi.org/10.1007/978-3-031-94016-3_5

Using (4.28) and subsequently (4.50), we arrive at the following equation

$$\frac{\partial \langle V_1 \rangle}{\partial x_1} + \frac{\partial \langle V_2 \rangle}{\partial x_2} + \frac{\partial \langle V_3 \rangle}{\partial x_3} = 0\,, \text{ or}$$
$$\frac{\partial \langle V_i \rangle}{\partial x_i} = 0 \tag{5.2}$$

where a repeated index implies summation over the full range of the index (Einstein's summation rule). Equation (5.2) is called the *mean continuity* equation. The dependent variables in this equation are the mean values of the velocity components, $\langle V_i \rangle$. This equation describes the application of conservation of mass in the turbulent flow field in the mean sense. Like the instantaneous continuity equation (3.10), the mean continuity equation (5.2), too, is a linear partial differential equation (PDE).

5.2 The Mean Momentum Equation

Next, we subject the momentum equation (3.11) to the mean operator

$$\left\langle \frac{\rho \partial V_i}{\partial t} + \rho V_k \frac{\partial V_i}{\partial x_k} \right\rangle = \left\langle -\frac{\partial p}{\partial x_i} + \mu \frac{\partial^2 V_i}{\partial x_k \partial x_k} \right\rangle \tag{5.3}$$

Employing (4.50) and (4.48) to the terms that involve time or spatial derivatives leads to

$$\rho \frac{\partial \langle V_i \rangle}{\partial t} + \left\langle \rho V_k \frac{\partial V_i}{\partial x_k} \right\rangle = -\frac{\partial \langle p \rangle}{\partial x_i} + \mu \frac{\partial^2 \langle V_i \rangle}{\partial x_k \partial x_k} \tag{5.4}$$

Since ρ and μ are constants, they commute across both the mean and the derivative operators.

The second term on the RHS Eq. (5.4) is the mean of the advection term. Using the instantaneous continuity equation (3.10), this terms is recast as

$$\left\langle \rho V_k \frac{\partial V_i}{\partial x_k} \right\rangle = \left\langle \frac{\partial (\rho V_k V_i)}{\partial x_k} - \rho \frac{\partial V_k}{\partial x_k} V_i \right\rangle = \left\langle \frac{\partial (\rho V_k V_i)}{\partial x_k} \right\rangle = \frac{\partial \langle \rho V_k V_i \rangle}{\partial x_k} \tag{5.5}$$

Using (4.36), (5.5) is expressed as

$$\frac{\partial \langle \rho V_k V_i \rangle}{\partial x_k} = \frac{\partial (\rho \langle V_k \rangle \langle V_i \rangle)}{\partial x_k} + \frac{\partial \left\langle \rho V_k' V_i' \right\rangle}{\partial x_k} \tag{5.6}$$

Further, employing the mean continuity equation (5.2), the RHS of (5.6) is simplified

$$\frac{\partial \langle \rho V_k V_i \rangle}{\partial x_k} = \rho \langle V_k \rangle \frac{\partial \langle V_i \rangle}{\partial x_k} + \rho \frac{\partial \langle V_k \rangle}{\partial x_k} \langle V_i \rangle + \frac{\partial \left\langle \rho V_k' V_i' \right\rangle}{\partial x_k}$$
$$\frac{\partial \langle \rho V_k V_i \rangle}{\partial x_k} = \rho \langle V_k \rangle \frac{\partial \langle V_i \rangle}{\partial x_k} + \frac{\partial \left\langle \rho V_k' V_i' \right\rangle}{\partial x_k} \tag{5.7}$$

Substituting (5.7) in (5.4) leads to

$$\rho \frac{\partial \langle V_i \rangle}{\partial t} + \rho \langle V_k \rangle \frac{\partial \langle V_i \rangle}{\partial x_k} = -\frac{\partial \langle p \rangle}{\partial x_i} + \mu \frac{\partial^2 \langle V_i \rangle}{\partial x_k \partial x_k} - \frac{\partial \left\langle \rho V_k^{'} V_i^{'} \right\rangle}{\partial x_k} \tag{5.8}$$

Equation (5.8) is called the *mean momentum equation*. The set of equations comprised of (5.2) and (5.8) is called the *mean Navier-Stokes equation* or the *Reynolds-averaged Navier-Stokes (RANS) equation set*. The *primary unknowns* of the mean Navier-Stokes equation set are the mean velocity components and mean pressure: $\langle V_i \rangle$ and $\langle p \rangle$. Further, the process of subjecting the non-linear advection term to the mean operator (5.5) has led to the emergence of six "new" or *secondary* unknowns $\left\langle V_1^{'} V_1^{'} \right\rangle, \left\langle V_1^{'} V_2^{'} \right\rangle, \left\langle V_1^{'} V_3^{'} \right\rangle, \left\langle V_2^{'} V_2^{'} \right\rangle, \left\langle V_2^{'} V_3^{'} \right\rangle$ and $\left\langle V_3^{'} V_3^{'} \right\rangle$. The six new unknowns on the RHS of (5.8) are related to a symmetric second-order tensor which is called the *Reynolds stress tensor* ($\underline{R}$):

$$\underline{R} = -\left\langle \rho V_i^{'} V_j^{'} \right\rangle \hat{e}_i \hat{e}_j \tag{5.9}$$

Like the instantaneous momentum equations (3.11), the mean momentum equation (5.8) is non-linear PDE. The non-linearity arises because of the advection term (second term on the LHS of 5.8).

Equations (5.2) and (5.8) have been presented using a frame-fixed Cartesian coordinate system as the working coordinate system. In its coordinate-system independent form, the RANS equation set is expressed as

$$\begin{gathered} \underline{\nabla} \cdot \langle \underline{V} \rangle = 0 \\ \rho \frac{\partial \langle \underline{V} \rangle}{\partial t} + \rho \underline{V} \cdot (\underline{\nabla} \langle \underline{V} \rangle) = -\underline{\nabla} \langle p \rangle + \mu \nabla^2 \langle \underline{V} \rangle + \underline{\nabla} \cdot \underline{R} \end{gathered} \tag{5.10}$$

5.3 The Turbulence Closure Problem

The instantaneous Navier-Stokes equation set (3.10)–(3.11) has four scalar PDEs in four unknown scalars (V_i and p). Thus, it is *mathematically closed*. The mean Navier-Stokes equation set, too, has four scalar PDEs (5.10). However, it involves 10 unknown scalars ($\langle V_1 \rangle, \langle V_2 \rangle, \langle V_3 \rangle, \langle p \rangle, \left\langle V_1^{'} V_1^{'} \right\rangle, \left\langle V_1^{'} V_2^{'} \right\rangle, \left\langle V_1^{'} V_3^{'} \right\rangle, \left\langle V_2^{'} V_2^{'} \right\rangle, \left\langle V_2^{'} V_3^{'} \right\rangle$ and $\left\langle V_3^{'} V_3^{'} \right\rangle$). Thus, the mean Navier-Stokes equation set is *mathematically unclosed*. This mismatch in the number of variables and the number of available equations involving these variables is called the *turbulence closure problem*. Thus, the mean Navier-Stokes set in its current form, even though being exact, is not complete enough to lead us to any solution of the mean flow field. To make any further progress towards obtaining a solution of the mean flow field with these equations, we must have additional equations that describe the variation of the six components of the Reynolds stress tensor that appear in (5.8).

5.4 The Reynolds Stress Tensor

The SI units of the Reynolds stress tensor (5.9) are the same as the SI units of the instantaneous stress tensor (Nm^{-2}). Indeed, we can show that the Reynolds stress arises due to the flux of fluctuating momentum vector caused by the fluctuating velocity vector across an (imaginary) surface inside the bulk of the fluid.

We consider a bulk of fluid, as shown in Fig. 5.1. Further, we consider an imaginary plane perpendicular to the direction of the unit vector $\hat{e}_1$ which passes through the location P which has its coordinates as (x_1, x_2, x_3) (Fig. 5.1). This plane divides the fluid bulk into two parts. We focus on the left part and identify this mass of the fluid as our system (Fig. 5.2). The outward normal unit vector for this exposed surface of the left bulk of the fluid is $\hat{e}_1$. We focus our attention on a small area ΔA (shown in Fig. 5.2) on this exposed surface, such that the centroid of this area coincides with location P (x_1, x_2, x_3). Over a unit time, the fluctuating velocity component V_1'

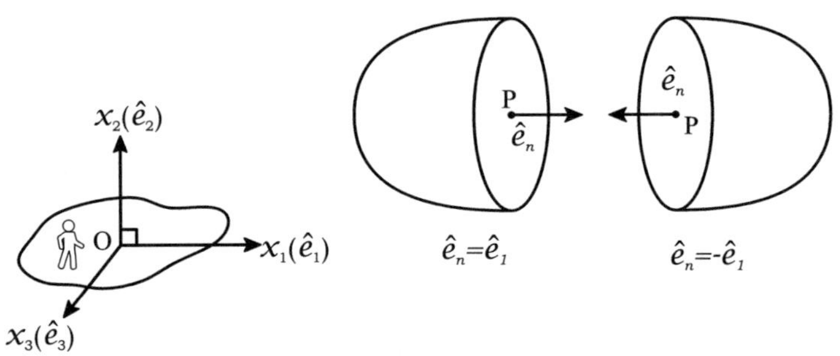

Fig. 5.1 Two parts of the fluid bulk created by an imaginary plane which is perpendicular to $\hat{e}_1$. The plane passes through location P. The two parts are shown separated and displaced only for the purpose of illustration

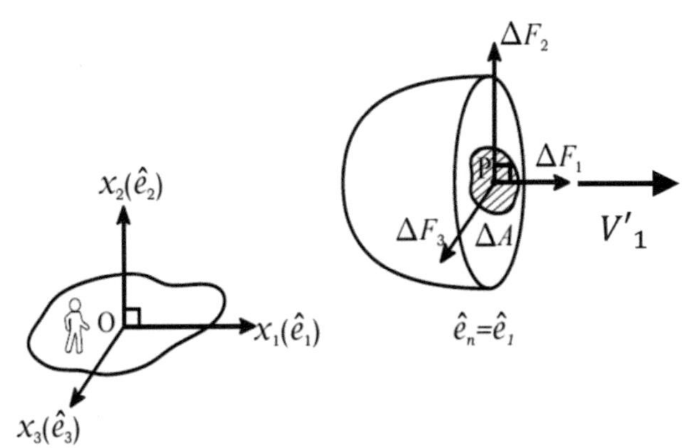

Fig. 5.2 Loss of fluctuating fluid momentum per unit time caused by the fluctuating velocity vector component V_1'

causes a certain amount of mass (Δm) to leave (per unit time) the left part of the fluid bulk through area ΔA.

$$\Delta m = \rho V_1{}' \Delta A \tag{5.11}$$

where V_1' is the fluctuating component of the velocity vector along $\hat{e}_1$ at time t and at the centroid of the small area ΔA. This leaving mass, in turn, causes, loss (per unit time) of a certain amount of fluctuating momentum vector, as well. This transport of the fluctuating momentum vector (denoted by $\Delta \underline{M}$ with SI units as $kgms^{-2}$) is represented as

$$\begin{aligned} \Delta \underline{M} &= \Delta m \left(V_1' \hat{e}_1 + V_2' \hat{e}_2 + V_3' \hat{e}_3 \right) \\ &= \rho V_1{}' \Delta A \left(V_1' \hat{e}_1 + V_2' \hat{e}_2 + V_3' \hat{e}_3 \right) \end{aligned} \tag{5.12}$$

The expected value of this momentum loss from the left bulk of the fluid per unit time can equivalently be perceived as a net external force vector $\underline{F}$ acting along $-\hat{e}_1$ direction on the left part of the fluid bulk.

$$\underline{F} = \left\langle -\rho V_1{}' \Delta A \left(V_1' \hat{e}_1 + V_2' \hat{e}_2 + V_3' \hat{e}_3 \right) \right\rangle \tag{5.13}$$

In the limit of $\Delta A \to 0$, the corresponding force components per unit area tend to match the definitions of the scalar components of the Reynolds stress tensor (5.9)

$$\lim_{\Delta A \to 0} \frac{\underline{F} \cdot \hat{e}_1}{\Delta A} = \lim_{\Delta A \to 0} \frac{\left\langle -\rho V_1{}' \Delta A V_1' \right\rangle}{\Delta A} = \left\langle -\rho V_1{}' V_1' \right\rangle = R_{11} \tag{5.14}$$

$$\lim_{\Delta A \to 0} \frac{\underline{F} \cdot \hat{e}_2}{\Delta A} = \lim_{\Delta A \to 0} \frac{\left\langle -\rho V_1{}' \Delta A V_2' \right\rangle}{\Delta A} = \left\langle -\rho V_1{}' V_2' \right\rangle = R_{12} \tag{5.15}$$

$$\lim_{\Delta A \to 0} \frac{\underline{F} \cdot \hat{e}_3}{\Delta A} = \lim_{\Delta A \to 0} \frac{\left\langle -\rho V_1{}' \Delta A V_3' \right\rangle}{\Delta A} = \left\langle -\rho V_1{}' V_3' \right\rangle = R_{13} \tag{5.16}$$

Similarly, we can demonstrate the origin of other scalar components of the Reynolds stress tensor by dividing the fluid bulk at location P using planes with the appropriate unit normal vectors ($\hat{e}_2$ or $\hat{e}_3$). The three components $\left\langle -\rho V_1' V_1' \right\rangle$, $\left\langle -\rho V_2' V_2' \right\rangle$ and $\left\langle -\rho V_3' V_3' \right\rangle$ are also called the *normal* Reynolds stress components. In contrast, the three components $\left\langle -\rho V_1' V_2' \right\rangle$, $\left\langle -\rho V_2' V_3' \right\rangle$ and $\left\langle -\rho V_3' V_1' \right\rangle$ are called the *shear* Reynolds stress components.

In an instantaneous turbulent flow field of a Newtonian fluid, the instantaneous stress tensor, which is an Eulerian variable (denoted by $\underline{\sigma}$), consists of two parts: the instantaneous pressure stress tensor and the instantaneous viscous stress tensor

$$\underline{\sigma} = -p \underline{I} + \underline{\tau} \tag{5.17}$$

where $\underline{I}$ denotes the identity tensor of order two, and

$$\underline{\tau} = 2\mu\underline{S} \tag{5.18}$$

The symbol $\underline{S}$ denotes the instantaneous strain-rate tensor. In the absence of body force, it is the instantaneous stress tensor which causes the instantaneous acceleration of a fluid particle.

$$\rho\frac{\partial \underline{V}}{\partial t} + \rho\underline{V}\cdot\left(\underline{\nabla}\underline{V}\right) = \underline{\nabla}\cdot\underline{\sigma} \tag{5.19}$$

In contrast, in the mean description of a turbulent flow, the effective stress tensor (we use the symbol $\underline{\sigma}^{\mathrm{mean}}$ to denote this quantity) that causes a fluid particle to have its mean acceleration is the sum of the mean of the instantaneous stress tensor and the Reynolds stress tensor.

$$\underline{\sigma}^{\mathrm{mean}} = \left\langle\underline{\sigma}\right\rangle + \underline{R} = -\left\langle p\right\rangle\underline{I} + \left\langle\underline{\tau}\right\rangle + \underline{R} \tag{5.20}$$

where

$$\left\langle\underline{\tau}\right\rangle = \left\langle 2\mu\underline{S}\right\rangle = 2\mu\left\langle\underline{S}\right\rangle = 2\mu\frac{\left(\underline{\nabla}\left\langle\underline{V}\right\rangle\right) + \left(\underline{\nabla}\left\langle\underline{V}\right\rangle\right)^T}{2} \tag{5.21}$$

where $\underline{R}$ is the Reynold stress tensor as defined in (5.9). Using the new symbol proposed in (5.20), the mean momentum equation (5.10) can be expressed as

$$\rho\frac{\partial\left\langle\underline{V}\right\rangle}{\partial t} + \rho(\left\langle\underline{V}\right\rangle\cdot\underline{\nabla})\left\langle\underline{V}\right\rangle = \underline{\nabla}\cdot\underline{\sigma}^{\mathrm{mean}} \tag{5.22}$$

Note that $\underline{\sigma}^{\mathrm{mean}}$ is not the same as $\left\langle\underline{\sigma}\right\rangle$.

5.5 Transport Equation of the Reynolds Stress Tensor

In Sect. 5.3, we identified the Reynolds stress tensor as the root cause of the turbulence closure problem. Addressing the turbulence closure problem requires additional equations that can describe the evolution of the Reynolds stress tensor, ensuring that the total number of unknowns matches the number of governing equations. Thus, motivated by this requirement, we now wish to derive the exact governing equations of the components of the Reynolds stress tensor. The instantaneous momentum equation (3.11), along with the mean momentum equation (5.8), provide the starting point of this derivation.

We first express the velocity and pressure variables in terms of the corresponding mean and fluctuating parts:

$$V_i = \langle V_i\rangle + V_i^{'},\ \text{and}\ p = \langle p\rangle + p^{'} \tag{5.23}$$

Thus, the instantaneous momentum equation (3.11) is expressed as

$$\frac{\partial\left(\langle V_i\rangle+V_i^{'}\right)}{\partial t}+\left(\langle V_k\rangle+V_k^{'}\right)\frac{\partial\left(\langle V_i\rangle+V_i^{'}\right)}{\partial x_k}= \\ -\frac{1}{\rho}\frac{\partial\left(\langle p\rangle+p^{'}\right)}{\partial x_i}+\nu\frac{\partial^2\left(\langle V_i\rangle+V_i^{'}\right)}{\partial x_k\partial x_k} \tag{5.24}$$

Further, subtracting the mean momentum equation (5.8) from (5.24) leads to the governing equation of $V_i^{'}$ as follows:

$$\frac{\partial V_i^{'}}{\partial t}+\langle V_q\rangle\frac{\partial V_i^{'}}{\partial x_q}=-V_q^{'}\frac{\partial V_i^{'}}{\partial x_q}-V_q^{'}\frac{\partial\langle V_i\rangle}{\partial x_q}-\frac{1}{\rho}\frac{\partial p^{'}}{\partial x_i}+\nu\frac{\partial^2 V_i^{'}}{\partial x_q\partial x_q} \\ +\frac{\partial\left\langle V_i^{'}V_q^{'}\right\rangle}{\partial x_q} \tag{5.25}$$

We multiply the scalar equation (5.25) by $V_j^{'}$ resulting into the following equation

$$V_j^{'}\frac{\partial V_i^{'}}{\partial t}+V_j^{'}\langle V_q\rangle\frac{\partial V_i^{'}}{\partial x_q}=-V_j^{'}V_q^{'}\frac{\partial V_i^{'}}{\partial x_q}-V_j^{'}V_q^{'}\frac{\partial\langle V_i\rangle}{\partial x_q}-V_j^{'}\frac{1}{\rho}\frac{\partial p^{'}}{\partial x_i} \\ +\nu V_j^{'}\frac{\partial^2 V_i^{'}}{\partial x_q\partial x_q}+V_j^{'}\frac{\partial\left\langle V_i^{'}V_q^{'}\right\rangle}{\partial x_q} \tag{5.26}$$

Changing the indices i's to j's and j's and i's in (5.26) results into another equation, which is independent of (5.26)

$$V_i^{'}\frac{\partial V_j^{'}}{\partial t}+V_i^{'}\langle V_q\rangle\frac{\partial V_j^{'}}{\partial x_q}=-V_i^{'}V_q^{'}\frac{\partial V_j^{'}}{\partial x_q}-V_i^{'}V_q^{'}\frac{\partial\langle V_j\rangle}{\partial x_q}-V_i^{'}\frac{1}{\rho}\frac{\partial p^{'}}{\partial x_j} \\ +\nu V_i^{'}\frac{\partial^2 V_j^{'}}{\partial x_q\partial x_q}+V_i^{'}\frac{\partial\left\langle V_j^{'}V_q^{'}\right\rangle}{\partial x_q} \tag{5.27}$$

Next, we add (5.26) and (5.27), subsequent similar terms from the two parent equations (Eqs. (5.26) and (5.27)) are grouped together for further algebraic simplifications.

$$\left[V_i^{'}\frac{\partial V_j^{'}}{\partial t}+V_j^{'}\frac{\partial V_i^{'}}{\partial t}\right]+\left[V_i^{'}\langle V_q\rangle\frac{\partial V_j^{'}}{\partial x_q}+V_j^{'}\langle V_q\rangle\frac{\partial V_i^{'}}{\partial x_q}\right] \\ =\left[-V_i^{'}V_q^{'}\frac{\partial V_j^{'}}{\partial x_q}-V_j^{'}V_q^{'}\frac{\partial V_i^{'}}{\partial x_q}\right]+\left[-V_i^{'}V_q^{'}\frac{\partial\langle V_j\rangle}{\partial x_q}-V_j^{'}V_q^{'}\frac{\partial\langle V_i\rangle}{\partial x_q}\right]$$

$$+\left[-V_i'\frac{1}{\rho}\frac{\partial p'}{\partial x_j}-V_j'\frac{1}{\rho}\frac{\partial p'}{\partial x_i}\right]+\left[V_i'\frac{\partial\left\langle V_j'V_q'\right\rangle}{\partial x_q}+V_j'\frac{\partial\left\langle V_i'V_q'\right\rangle}{\partial x_q}\right]$$
$$+\left[\nu V_i'\frac{\partial^2 V_j'}{\partial x_q\partial x_q}+\nu V_j'\frac{\partial^2 V_i'}{\partial x_q\partial x_q}\right] \tag{5.28}$$

Next, we take the mean of the entire Eq. (5.28). Subsequently, using the distributive property of the mean operator over a sum of random variables (4.28), we arrive at

$$\underbrace{\left\langle V_i'\frac{\partial V_j'}{\partial t}+V_j'\frac{\partial V_i'}{\partial t}\right\rangle}_{I}+\underbrace{\left\langle V_i'\left\langle V_q\right\rangle\frac{\partial V_j'}{\partial x_q}+V_j'\left\langle V_q\right\rangle\frac{\partial V_i'}{\partial x_q}\right\rangle}_{II}$$
$$=\underbrace{\left\langle -V_i'V_q'\frac{\partial V_j'}{\partial x_q}-V_j'V_q'\frac{\partial V_i'}{\partial x_q}\right\rangle}_{III}+\underbrace{\left\langle -V_i'V_q'\frac{\partial\langle V_j\rangle}{\partial x_q}-V_j'V_q'\frac{\partial\langle V_i\rangle}{\partial x_q}\right\rangle}_{IV}$$
$$+\underbrace{\left\langle -V_i'\frac{1}{\rho}\frac{\partial p'}{\partial x_j}-V_j'\frac{1}{\rho}\frac{\partial p'}{\partial x_i}\right\rangle}_{V}+\underbrace{\left\langle V_i'\frac{\partial\left\langle V_j'V_q'\right\rangle}{\partial x_q}+V_j'\frac{\partial\left\langle V_i'V_q'\right\rangle}{\partial x_q}\right\rangle}_{VI}$$
$$+\underbrace{\left\langle \nu V_i'\frac{\partial^2 V_j'}{\partial x_q\partial x_q}+\nu V_j'\frac{\partial^2 V_i'}{\partial x_q\partial x_q}\right\rangle}_{VII} \tag{5.29}$$

Further, we simplify terms I–VII, individually. Various properties of the mean operator (Eqs. 4.22, 4.28, 4.47 and 4.50) are employed for the simplification process.

$$\underbrace{\left\langle V_i'\frac{\partial V_j'}{\partial t}+V_j'\frac{\partial V_i'}{\partial t}\right\rangle}_{I}=\left\langle\frac{\partial\left(V_i'V_j'\right)}{\partial t}\right\rangle=\frac{\partial\left\langle V_i'V_j'\right\rangle}{\partial t} \tag{5.30}$$

Term II simplifies to

$$\underbrace{\left\langle V_i'\left\langle V_q\right\rangle\frac{\partial V_j'}{\partial x_q}+V_j'\left\langle V_q\right\rangle\frac{\partial V_i'}{\partial x_q}\right\rangle}_{II}=\left\langle\left\langle V_q\right\rangle\frac{\partial\left(V_i'V_j'\right)}{\partial x_q}\right\rangle$$
$$=\left\langle V_q\right\rangle\frac{\partial\left\langle V_i'V_j'\right\rangle}{\partial x_q} \tag{5.31}$$

Term III simplifies to

$$\underbrace{\left\langle -V_i'V_q'\frac{\partial V_j'}{\partial x_q} - V_j'V_q'\frac{\partial V_i'}{\partial x_q}\right\rangle}_{III}$$
$$= \left\langle -\frac{\partial\left(V_i'V_j'V_q'\right)}{\partial x_q} + V_i'V_j'\frac{\partial V_q'}{\partial x_q}\right\rangle \tag{5.32}$$

To further simplify the last expression, we first subtract (5.2) from (3.10) and obtain the *fluctuating continuity equation*

$$\frac{\partial V_q'}{\partial x_q} = 0. \tag{5.33}$$

Next, using (5.33) in (5.32), simplifies Term III to

$$\underbrace{\left\langle -V_i'V_q'\frac{\partial V_j'}{\partial x_q} - V_j'V_q'\frac{\partial V_i'}{\partial x_q}\right\rangle}_{III} = \left\langle -\frac{\partial\left(V_i'V_j'V_q'\right)}{\partial x_q} + V_i'V_j'\frac{\partial V_q'}{\partial x_q}\right\rangle$$
$$= \left\langle -\frac{\partial\left(V_i'V_j'V_q'\right)}{\partial x_q}\right\rangle = -\frac{\partial\left\langle V_i'V_j'V_q'\right\rangle}{\partial x_q} \tag{5.34}$$

Term IV simplifies to

$$\underbrace{\left\langle -V_i'V_q'\frac{\partial\langle V_j\rangle}{\partial x_q} - V_j'V_q'\frac{\partial\langle V_i\rangle}{\partial x_q}\right\rangle}_{IV}$$
$$= -\left\langle V_i'V_q'\right\rangle\frac{\partial\langle V_j\rangle}{\partial x_q} - \left\langle V_j'V_q'\right\rangle\frac{\partial\langle V_i\rangle}{\partial x_q} \tag{5.35}$$

Term V is split into two different groups of terms (explanation to be provided later in Chap. 9).

$$\underbrace{\left\langle -V_i^{'}\frac{1}{\rho}\frac{\partial p^{'}}{\partial x_j} - V_j^{'}\frac{1}{\rho}\frac{\partial p^{'}}{\partial x_i}\right\rangle}_{V}$$

$$= \left\langle \left(-\frac{1}{\rho}\frac{\partial \left(p^{'}V_i^{'}\right)}{\partial x_j} + \frac{p^{'}}{\rho}\frac{\partial V_i^{'}}{\partial x_j} \right) + \left(-\frac{1}{\rho}\frac{\partial \left(p^{'}V_j^{'}\right)}{\partial x_i} + \frac{p^{'}}{\rho}\frac{\partial V_j^{'}}{\partial x_i} \right) \right\rangle$$

$$= \left\langle -\frac{1}{\rho}\frac{\partial \left(p^{'}V_i^{'}\right)}{\partial x_j} + \frac{p^{'}}{\rho}\frac{\partial V_i^{'}}{\partial x_j} \right\rangle + \left\langle -\frac{1}{\rho}\frac{\partial \left(p^{'}V_j^{'}\right)}{\partial x_i} + \frac{p^{'}}{\rho}\frac{\partial V_j^{'}}{\partial x_i} \right\rangle$$

$$= -\frac{1}{\rho}\frac{\partial \left\langle p^{'}V_i^{'}\right\rangle}{\partial x_j} + \left\langle \frac{p^{'}}{\rho}\frac{\partial V_i^{'}}{\partial x_j} \right\rangle - \frac{1}{\rho}\frac{\partial \left\langle p^{'}V_j^{'}\right\rangle}{\partial x_i} + \left\langle \frac{p^{'}}{\rho}\frac{\partial V_j^{'}}{\partial x_i} \right\rangle$$

$$= -\frac{1}{\rho}\left(\frac{\partial \left\langle p^{'}V_i^{'}\right\rangle}{\partial x_j} + \frac{\partial \left\langle p^{'}V_j^{'}\right\rangle}{\partial x_i} \right) + \frac{1}{\rho}\left\langle p^{'}\left(\frac{\partial V_j^{'}}{\partial x_i} + \frac{\partial V_i^{'}}{\partial x_j} \right) \right\rangle$$

$$= -\frac{1}{\rho}\left(\frac{\partial \left\langle p^{'}V_i^{'}\right\rangle}{\partial x_j} + \frac{\partial \left\langle p^{'}V_j^{'}\right\rangle}{\partial x_i} \right) + \frac{2}{\rho}\left\langle p^{'}s_{ij}^{'}\right\rangle \tag{5.36}$$

where $s_{ij}^{'}$ represents the (ij)th component of the fluctuating strain-rate tensor (2.21)

$$s_{ij}^{'} = \frac{1}{2}\left(\frac{\partial V_j^{'}}{\partial x_i} + \frac{\partial V_i^{'}}{\partial x_j} \right) \tag{5.37}$$

The first term on the RHS of (5.36) can be altered using the Kronecker delta symbols to cast Term V

$$\underbrace{\left\langle -V_i^{'}\frac{1}{\rho}\frac{\partial p^{'}}{\partial x_j} - V_j^{'}\frac{1}{\rho}\frac{\partial p^{'}}{\partial x_i}\right\rangle}_{V} =$$

$$-\frac{1}{\rho}\left(\frac{\partial \left\langle p^{'}V_i^{'}\right\rangle}{\partial x_q}\delta_{jq} + \frac{\partial \left\langle p^{'}V_j^{'}\right\rangle}{\partial x_q}\delta_{iq} \right) + \frac{2}{\rho}\left\langle p^{'}s_{ij}^{'}\right\rangle \tag{5.38}$$

Term VI simplifies to

$$\underbrace{\left\langle V_i^{'} \frac{\partial \left\langle V_j^{'} V_q^{'} \right\rangle}{\partial x_q} + V_j^{'} \frac{\partial \left\langle V_i^{'} V_q^{'} \right\rangle}{\partial x_q} \right\rangle}_{VI}$$

$$= \left\langle V_i^{'} \right\rangle \frac{\partial \left\langle V_j^{'} V_q^{'} \right\rangle}{\partial x_q} + \left\langle V_j^{'} \right\rangle \frac{\partial \left\langle V_i^{'} V_q^{'} \right\rangle}{\partial x_q} = 0. \tag{5.39}$$

Term VII is expressed as

$$\underbrace{\left\langle \nu V_i^{'} \frac{\partial^2 V_j^{'}}{\partial x_q \partial x_q} + \nu V_j^{'} \frac{\partial^2 V_i^{'}}{\partial x_q \partial x_q} \right\rangle}_{VII} = \nu \left\langle \left(V_i^{'} \frac{\partial^2 V_j^{'}}{\partial x_q \partial x_q} \right) + \left(V_j^{'} \frac{\partial^2 V_i^{'}}{\partial x_q \partial x_q} \right) \right\rangle$$

$$= \nu \left\langle \left(\frac{\partial}{\partial x_q} \left[\frac{\partial V_j^{'}}{\partial x_q} V_i^{'} \right] - \frac{\partial V_i^{'}}{\partial x_q} \frac{\partial V_j^{'}}{\partial x_q} \right) + \left(\frac{\partial}{\partial x_q} \left[\frac{\partial V_i^{'}}{\partial x_q} V_j^{'} \right] - \frac{\partial V_j^{'}}{\partial x_q} \frac{\partial V_i^{'}}{\partial x_q} \right) \right\rangle$$

$$= \nu \left\langle \left(\frac{\partial}{\partial x_q} \left[\frac{\partial V_j^{'}}{\partial x_q} V_i^{'} \right] + \frac{\partial}{\partial x_q} \left[\frac{\partial V_i^{'}}{\partial x_q} V_j^{'} \right] \right) + \left(- \frac{\partial V_i^{'}}{\partial x_q} \frac{\partial V_j^{'}}{\partial x_q} - \frac{\partial V_j^{'}}{\partial x_q} \frac{\partial V_i^{'}}{\partial x_q} \right) \right\rangle$$

$$= \nu \left\langle \frac{\partial}{\partial x_q} \left(\frac{\partial V_j^{'}}{\partial x_q} V_i^{'} + \frac{\partial V_i^{'}}{\partial x_q} V_j^{'} \right) - 2 \frac{\partial V_i^{'}}{\partial x_q} \frac{\partial V_j^{'}}{\partial x_q} \right\rangle$$

$$= \nu \left\langle \frac{\partial}{\partial x_q} \left(\frac{\partial \left[V_j^{'} V_i^{'} \right]}{\partial x_q} \right) - 2 \frac{\partial V_i^{'}}{\partial x_q} \frac{\partial V_j^{'}}{\partial x_q} \right\rangle$$

$$= \nu \left\langle \frac{\partial^2}{\partial x_q \partial x_q} \left(V_j^{'} V_i^{'} \right) - 2 \frac{\partial V_i^{'}}{\partial x_q} \frac{\partial V_j^{'}}{\partial x_q} \right\rangle$$

$$= \nu \frac{\partial^2 \left\langle V_j^{'} V_i^{'} \right\rangle}{\partial x_q \partial x_q} - 2\nu \left\langle \frac{\partial V_i^{'}}{\partial x_q} \frac{\partial V_j^{'}}{\partial x_q} \right\rangle \tag{5.40}$$

Substituting (5.30), (5.31), (5.34), (5.35), (5.38), (5.39) and (5.40) in (5.29) and multiplying the entire equation by ρ we arrive at the following equation

$$\begin{aligned}
&\frac{\partial \left\langle \rho V_i^{'} V_j^{'} \right\rangle}{\partial t} + \left\langle V_q \right\rangle \frac{\partial \left\langle \rho V_i^{'} V_j^{'} \right\rangle}{\partial x_q} \\
&= \left(- \left\langle \rho V_i^{'} V_q^{'} \right\rangle \frac{\partial \langle V_j \rangle}{\partial x_q} - \left\langle \rho V_j^{'} V_q^{'} \right\rangle \frac{\partial \langle V_i \rangle}{\partial x_q} \right) - 2\mu \left\langle \frac{\partial V_i^{'}}{\partial x_q} \frac{\partial V_j^{'}}{\partial x_q} \right\rangle \\
&+ 2 \left\langle p^{'} s_{ij}^{'} \right\rangle + \nu \frac{\partial^2 \left\langle \rho V_j^{'} V_i^{'} \right\rangle}{\partial x_q \partial x_q} \\
&+ \left(- \frac{\partial \left\langle p^{'} V_i^{'} \right\rangle}{\partial x_j} - \frac{\partial \left\langle p^{'} V_j^{'} \right\rangle}{\partial x_i} \right) + \frac{\partial}{\partial x_q} \left\langle - \rho V_q^{'} V_i^{'} V_j^{'} \right\rangle
\end{aligned} \tag{5.41}$$

Using the definition (5.9), Eq. (5.41) is cast as the transport equation of R_{ij}.

$$\begin{aligned}
\frac{\partial \underline{R}}{\partial t} + \langle V_k \rangle \frac{\partial \underline{R}}{\partial x_k} &= -R_{iq} \frac{\partial \langle V_j \rangle}{\partial x_i} - R_{jq} \frac{\partial \langle V_i \rangle}{\partial x_q} + 2\mu \left\langle \frac{\partial V_i^{'}}{\partial x_q} \frac{\partial V_j^{'}}{\partial x_q} \right\rangle \\
&- 2 \left\langle p^{'} s_{ij}^{'} \right\rangle + \nu \frac{\partial^2 R_{ij}}{\partial x_q \partial x_q} \\
&+ \left(\frac{\partial \left\langle p^{'} V_i^{'} \right\rangle}{\partial x_j} + \frac{\partial \left\langle p^{'} V_j^{'} \right\rangle}{\partial x_i} \right) + \frac{\partial}{\partial x_q} \left\langle \rho V_q^{'} V_i^{'} V_j^{'} \right\rangle
\end{aligned} \tag{5.42}$$

Symbolically (5.42) is expressed as

$$\frac{\partial R_{ij}}{\partial t} + \left\langle V_q \right\rangle \frac{\partial R_{ij}}{\partial x_q} = \mathcal{P}_{ij} + \epsilon_{ij} - \Pi_{ij} + \Gamma_{ij} - \mathcal{T}_{ij} \tag{5.43}$$

where, $\mathcal{P}_{ij}$, ϵ_{ij}, Π_{ij}, Γ_{ij} and $\mathcal{T}_{ij}$ are the (ij)th components of the following tensors.

$$\underline{\mathcal{P}} = -\left(\underline{R} \cdot \left(\underline{\nabla} \left\langle \underline{V} \right\rangle\right)\right) - \left(\underline{R} \cdot \left(\underline{\nabla} \left\langle \underline{V} \right\rangle\right)\right)^T \tag{5.44}$$

$$\underline{\epsilon} = 2\mu \left\langle \left(\underline{\nabla}\, \underline{V}^{'}\right)^T \cdot \left(\underline{\nabla}\, \underline{V}^{'}\right) \right\rangle \tag{5.45}$$

$$\underline{\Pi} = 2 \left\langle p^{'} \underline{s}^{'} \right\rangle \tag{5.46}$$

$$\underline{\Gamma} = \nu \nabla^2 \left(\underline{R}\right) \tag{5.47}$$

$$\underline{\mathcal{T}} = -\left(\underline{\nabla}\left\langle p^{'}\underline{V}^{'}\right\rangle\right) - \left(\underline{\nabla}\left\langle p^{'}\underline{V}^{'}\right\rangle\right)^{T} - \underline{\nabla}\cdot\left\langle\rho\underline{V}^{'}\,\underline{V}^{'}\,\underline{V}^{'}\right\rangle \tag{5.48}$$

Accordingly, the corresponding coordinate system-independent form of (5.43) is

$$\frac{\partial \underline{R}}{\partial t} + (\langle\underline{V}\rangle\cdot\underline{\nabla})\underline{R} = \underline{\mathcal{P}} + \underline{\epsilon} - \underline{\Pi} + \underline{\Gamma} - \underline{\mathcal{T}} \tag{5.49}$$

In the literature of turbulent flows, a partial differential equation of the form

$$\frac{\partial\psi}{\partial t} + \left\langle V_q\right\rangle\frac{\partial\psi}{\partial x_q} = P_1 + P_2 + \ \dots\ + P_N, \tag{5.50}$$

is called the *mean transport equation* of ψ, and the terms P_1, P_2, ... and P_N appearing on the RHS of (5.50) are called the *processes* that influence the evolution of ψ in the flow field. On the other side of the equation (the left-hand side), the first term is called the *unsteady term*, and the second term is called the *advection process*. The two partial differentiation operators on the LHS of (5.50) are often combined in the following form.

$$\frac{\partial}{\partial t} + \left\langle V_q\right\rangle\frac{\partial}{\partial x_q} \tag{5.51}$$

Physically, this combined operator represents the rate of change in ϕ (the quantity on which it acts) following the local mean velocity $\left\langle\underline{V}(\underline{X}, t)\right\rangle$. In other words, the LHS of (5.50) represents the rate of change in the quantity ψ following the *mean motion* of the local fluid particle. This operator is called the *mean material derivative* operator. In accordance with (5.50), Eq. (5.49) is called the *mean transport equation* of the Reynolds stress tensor. For brevity, we refer to this Eq. (5.49) simply by the acronym RSTE.

Equation (5.49) shows that the rate of change of the Reynolds stress tensor following the mean velocity of the local fluid particle happens under the influence of five processes: $\underline{\mathcal{P}}$, $\underline{\epsilon}$, $\underline{\Pi}$, $\underline{\Gamma}$ and $\underline{\mathcal{T}}$. The tensor $\underline{\mathcal{P}}$ is called the *production process* of the Reynolds stress tensor. The tensor $\underline{\epsilon}$ is called the *dissipation process* of the Reynolds stress tensor. The tensor $\underline{\Pi}$ is called the *pressure-strain correlation process*. The tensor Γ represents the *molecular diffusion process* of the Reynolds stress tensor in the turbulent flow field. The tensor $\underline{\mathcal{T}}$ involves gradients of the covariance of pressure and velocity, along with the divergence of the triple correlation of fluctuating velocity components.

5.6 Turbulence Kinetic Energy

Some additional physical insights into Eq. (5.49) and the involved processes can be gained by focusing on the trace of this equation. The trace of (5.49) can be readily obtained by setting $j = i$ in (5.43).

$$\frac{\partial R_{ii}}{\partial t} + \langle V_q \rangle \frac{\partial R_{ii}}{\partial x_q} = \mathcal{P}_{ii} + \epsilon_{ii} - \Pi_{ii} + \Gamma_{ii} - \mathcal{T}_{ii} \tag{5.52}$$

The trace of the Reynolds stress tensor is related to *turbulence kinetic energy per unit mass of the fluid* (denoted by symbol k, and quite often called, in a more simple manner, as the *turbulence kinetic energy*):

$$k = \frac{\left\langle V_i^{'} V_i^{'} \right\rangle}{2} = -\frac{R_{ii}}{2\rho}. \tag{5.53}$$

Indeed, the turbulence kinetic energy has a clear physical meaning. It represents the mean of part of instantaneous kinetic energy contained in the fluctuating velocity field. This is illustrated by the following relationship.

$$\left\langle \frac{V_i V_i}{2} \right\rangle = \frac{\langle V_i \rangle \langle V_i \rangle}{2} + \frac{\left\langle V_i^{'} V_i^{'} \right\rangle}{2} = \frac{\langle V_i \rangle \langle V_i \rangle}{2} + k. \tag{5.54}$$

The LHS represents the *mean of kinetic energy per unit mass*. The first term on the RHS represents the *kinetic energy per unit mean associated with the mean velocity field*, and the second term on the RHS of (5.54) represents the *the turbulence kinetic energy per unit mass*.

Using (5.52) and (5.53), the transport equation of the turbulence kinetic energy is expressed as

$$\frac{\partial k}{\partial t} + \langle V_q \rangle \frac{\partial k}{\partial x_q} = -\frac{\mathcal{P}_{ii}}{2\rho} - \frac{\epsilon_{ii}}{2\rho} + \frac{\Pi_{ii}}{2\rho} - \frac{\Gamma_{ii}}{2\rho} + \frac{\mathcal{T}_{ii}}{2\rho} \tag{5.55}$$

The full algebraic expression of this equation is

$$\begin{aligned}\frac{\partial k}{\partial t} + \langle V_q \rangle \frac{\partial k}{\partial x_q} = &\underbrace{-\left\langle V_i^{'} V_j^{'} \right\rangle \frac{\partial \langle V_i \rangle}{\partial x_j}}_{I^{'}} \underbrace{- \nu \left\langle \frac{\partial V_i^{'}}{\partial x_q} \frac{\partial V_i^{'}}{\partial x_q} \right\rangle}_{II^{'}} + \underbrace{\frac{1}{\rho} \left\langle p^{'} s_{ii}^{'} \right\rangle}_{III^{'}} \\ &+ \underbrace{\nu \frac{\partial^2 k}{\partial x_q \partial x_q}}_{IV^{'}} - \underbrace{\frac{1}{\rho} \frac{\left\langle p^{'} V_i^{'} \right\rangle}{\partial x_i}}_{V^{'}} - \underbrace{\frac{\partial}{\partial x_q} \left\langle V_q^{'} \frac{V_i^{'} V_i^{'}}{2} \right\rangle}_{VI^{'}}\end{aligned} \tag{5.56}$$

The first term on the RHS of (5.56) $\left(I^{'} = -\left\langle V_i^{'} V_j^{'} \right\rangle \frac{\partial \langle V_i \rangle}{\partial x_j} \right)$ is called the *production rate of* k. It can be demonstrated that $-I^{'}$ appears in the mean transport equation of the turbulence kinetic energy associated with mean velocity ($\langle V_k \rangle \langle V_k \rangle /2$). The transport equation of $\langle V_k\rangle \langle V_k\rangle /2$ can be readily obtained by multiplying (5.8) by $\langle V_i\rangle$, and subsequently taking the trace of the resulting equation.

$$\begin{aligned}&\frac{\partial}{\partial t}\left(\frac{\langle V_i\rangle \langle V_i\rangle}{2}\right)+\left\langle V_q\right\rangle \frac{\partial}{\partial x_q}\left(\frac{\langle V_i\rangle \langle V_i\rangle}{2}\right)\\&=-\frac{1}{\rho}\frac{\partial \langle p\rangle}{\partial x_i}\langle V_i\rangle+\nu\frac{\partial^2 \langle V_i\rangle}{\partial x_k \partial x_k}\langle V_i\rangle-\frac{\partial}{\partial x_p}\left(\left\langle V_i^{'}V_p^{'}\right\rangle \langle V_i\rangle\right)\\&+\underbrace{\left\langle V_i^{'}V_p^{'}\right\rangle \frac{\partial \langle V_i\rangle}{\partial x_p}}_{-I^{'}}\end{aligned} \tag{5.57}$$

The presence of the same process $I^{'}$ in Eqs. (5.56) and (5.57) with opposite signs shows that $I^{'}$ is indeed the mechanism by which energy is exchanged between the kinetic energy associated with the mean velocity ($\langle V_i\rangle \langle V_i\rangle /2$), and the kinetic energy associated with the fluctuating field $\left(\left\langle V_i^{'}V_i^{'}\right\rangle /2\right)$ of the same local fluid particle.

The second term on the RHS of (5.56) is called the *dissipation rate of* k. It is often denoted by the symbol ϵ

$$\epsilon=\nu\left\langle \frac{\partial V_i^{'}}{\partial x_q}\frac{\partial V_i^{'}}{\partial x_q}\right\rangle \tag{5.58}$$

Since ϵ involves the mean of the *sum of the square of all nine components of the fluctuating velocity gradient tensor*, ϵ is a positive-definite quantity.

$$\epsilon \geq 0. \tag{5.59}$$

This realization leads to the conclusion, that the dissipation rate can never increase the turbulence kinetic energy associated with a fluid particle. Whenever the gradient of the fluctuating velocity field exists ($\epsilon \neq 0$), the dissipation rate decreases k.

Further, we can demonstrate that the dissipation rate process (ϵ) appears with a positive sign in the mean transport equation of the mean internal energy of the flow field. This equation can be derived by taking the mean of the instantaneous governing equation of the internal energy per unit mass (e), which is another Eulerian variable. The instantaneous equation of e for a *Newtonian* fluid with constant density and without any source term is [1]:

$$\frac{\partial e}{\partial t}+V_q\frac{\partial e}{\partial x_q}=2\nu S_{mn}S_{mn}+\frac{\kappa}{\rho}\frac{\partial^2 T}{\partial x_k \partial x_k}, \tag{5.60}$$

where $\underline{S}$ represents the instantaneous stain-rate tensor. The symbol κ, represents the thermal conductivity of the fluid medium. Taking the mean of (5.60) and using relevant properties of the mean operator, results into the following equations.

$$\left\langle \frac{\partial e}{\partial t} + V_q \frac{\partial e}{\partial x_q} \right\rangle = \left\langle 2\nu S_{mn} S_{mn} + \frac{\kappa}{\rho} \frac{\partial^2 T}{\partial x_k \partial x_k} \right\rangle$$

$$\frac{\partial \langle e \rangle}{\partial t} + \langle V_q \rangle \frac{\partial \langle e \rangle}{\partial x_q} + \left\langle V_q^{'} \frac{\partial e^{'}}{\partial x_q} \right\rangle = 2\nu \langle S_{mn} \rangle \langle S_{mn} \rangle + 2\nu \left\langle S_{mn}^{'} S_{mn}^{'} \right\rangle + \frac{\kappa}{\rho} \frac{\partial^2 \langle T \rangle}{\partial x_k \partial x_k} \tag{5.61}$$

Substituting

$$S_{mn}^{'} = \frac{1}{2} \left(\frac{\partial V_m^{'}}{\partial x_n} + \frac{\partial V_n^{'}}{\partial x_m} \right) \tag{5.62}$$

in (5.61) leads to

$$\frac{\partial \langle e \rangle}{\partial t} + \langle V_q \rangle \frac{\partial \langle e \rangle}{\partial x_q} = -\left\langle V_q^{'} \frac{\partial e^{'}}{\partial x_q} \right\rangle + 2\nu \langle S_{mn} \rangle \langle S_{mn} \rangle + \frac{\kappa}{\rho} \frac{\partial^2 \langle T \rangle}{\partial x_k \partial x_k} + \underbrace{\nu \left\langle \frac{\partial V_m^{'}}{\partial x_n} \frac{\partial V_m^{'}}{\partial x_n} \right\rangle}_{\epsilon} + \nu \left\langle \frac{\partial V_n^{'}}{\partial x_m} \frac{\partial V_m^{'}}{\partial x_n} \right\rangle \tag{5.63}$$

The simultaneous presence of ϵ on the RHS of (5.56) and on the RHS of (5.63) with opposite signs clearly shows that ϵ is the mechanism by which the turbulence kinetic energy is converted to the mean internal energy (heat) associated with the local fluid particle.

The third term on the RHS of (5.56) involves the covariance of pressure and strain-rate. The mean continuity equation (5.33) implies that there is no net effect of the pressure-strain correlation process on the evolution of k.

$$\frac{1}{\rho} \left\langle p^{'} S_{ii}^{'} \right\rangle = \frac{1}{\rho} \left\langle p^{'} S_{11}^{'} \right\rangle + \frac{1}{\rho} \left\langle p^{'} S_{22}^{'} \right\rangle + \frac{1}{\rho} \left\langle p^{'} S_{33}^{'} \right\rangle = 0. \tag{5.64}$$

However, it does not mean the three individual components $\frac{1}{\rho} \left\langle p^{'} S_{11}^{'} \right\rangle$, $\frac{1}{\rho} \left\langle p^{'} S_{22}^{'} \right\rangle$ and $\frac{1}{\rho} \left\langle p^{'} S_{33}^{'} \right\rangle$, are unimportant. These individual components, in general, can still cause the individual Reynolds stress components $\left\langle V_1^{'} V_1^{'} \right\rangle$, $\left\langle V_2^{'} V_2^{'} \right\rangle$ and $\left\langle V_3^{'} V_3^{'} \right\rangle$ to decrease/increase. For example, if $\frac{1}{\rho} \left\langle p^{'} S_{11}^{'} \right\rangle > 0$ (causing an increase in $\left\langle V_1^{'} V_1^{'} \right\rangle$ in accordance with (5.43)) as well as $\frac{1}{\rho} \left\langle p^{'} S_{22}^{'} \right\rangle > 0$ (causing an increase in $\left\langle V_2^{'} V_2^{'} \right\rangle$ in accordance with (5.43)), the constraint (5.64) guarantees that in such a situation:

$$\frac{1}{\rho} \left\langle p^{'} S_{33}^{'} \right\rangle = -\left(\frac{1}{\rho} \left\langle p^{'} S_{11}^{'} \right\rangle + \frac{1}{\rho} \left\langle p^{'} S_{22}^{'} \right\rangle \right) < 0 \tag{5.65}$$

and correspondingly, the pressure strain-correlation must cause a decrease in $\left\langle V_3' V_3' \right\rangle$. Thus, the apparent role of the pressure strain correlation tensor on k is to cause a *redistribution* of the relative share of turbulence kinetic energy contributed by the three orthogonal fluctuating velocity components.

Term IV' on the RHS of (5.56) is the *molecular diffusion* process of k. In contrast to the role of ϵ, which causes a *local* conversion of turbulence kinetic energy into heat, the molecular process causes a spatial redistribution of k in the flow field from a region where k is higher to those where k is lower.

It can be demonstrated (detailed derivation not included here) that the term $-\frac{1}{\rho}\frac{\partial\left\langle p' V_i' \right\rangle}{\partial x_i}$ is related to the expectation of the rate of work done per unit mass by the fluctuating pressure forces while the local fluid particle undergoes displacement due to the fluctuating velocity vector ($\underline{V}'$).

To explain the physical meaning of Term VI', we first define a quantity called the *fluctuating kinetic energy per unit mass*. This is defined as $V_q' V_q'/2$. Next, we refer to Fig. 5.1. Over a unit time, the fluctuating velocity component V_1' causes a certain amount of mass (Δm) to be transported across the small area ΔA.

$$\Delta m = \rho V_1{}' \Delta A \tag{5.66}$$

Associated with this transported mass, the amount of fluctuating kinetic energy that is transported is $\Delta m V_k' V_k'/2$. Thus, the amount of fluctuating kinetic energy being transported per unit area per unit time is $\rho V_1{}' V_k' V_k'/2$. We call this quantity the *flux of fluctuating kinetic energy caused by the fluctuating velocity component* V_1'. Similarly, if we consider a small area ΔA with unit normal along the $\hat{e}_2$ direction (or along the $\hat{e}_3$ direction), we get the corresponding flux of the fluctuating kinetic energy caused by the V_2' component (or the V_3' component) as $\rho V_2{}' V_k' V_k'/2$ (or $\rho V_3{}' V_k' V_k'/2$). Thus, we define the ith component of a relevant flux vector ($\underline{Q}$) as

$$Q_i = \rho V_i' \frac{V_p' V_p'}{2} \tag{5.67}$$

The Term VI' can now be expressed in terms of the $\underline{Q}$ vector.

$$\underbrace{-\frac{\partial}{\partial x_q}\left\langle V_q' \frac{V_p' V_p'}{2} \right\rangle}_{VI'} = -\frac{1}{\rho}\frac{\partial \langle Q_q \rangle}{\partial x_q} = -\frac{1}{\rho}\underline{\nabla} \cdot \langle \underline{Q} \rangle \tag{5.68}$$

Thus, Term VI' represents the gradient of the flux of the fluctuating turbulence kinetic energy caused by the transporting action of the fluctuating velocity vector itself.

Earlier in Chap. 4 we defined a *statistically homogeneous flow field*, which has the expected values of all variables and their moments spatially independent. There exists another class of turbulent flows wherein, even though the expected value of the primary flow variables (velocity and pressure) may not be spatially independent, however, the expected values of all the products of the fluctuating quantities are still spatially independent

$$\frac{\partial \left\langle \phi_1^{'} \phi_2^{'} ... \right\rangle}{\partial x_i} = 0, \tag{5.69}$$

where $\phi_1^{'}$, $\phi_2^{'}$, etc. represent the fluctuating parts of various flow variables. Such a turbulent flow is called *homogeneous turbulence*. In such a flow field, the following processes appearing in the transport equation of the Reynolds stress tensor (5.41) are zero:

$$\begin{aligned}
&\left\langle V_q \right\rangle \frac{\partial \left\langle \rho V_i^{'} V_j^{'} \right\rangle}{\partial x_q} = 0 \\
&\nu \frac{\partial^2 \left\langle \rho V_j^{'} V_i^{'} \right\rangle}{\partial x_q \partial x_q} = 0 \\
&\left(-\frac{\partial \left\langle p^{'} V_i^{'} \right\rangle}{\partial x_j} - \frac{\partial \left\langle p^{'} V_j^{'} \right\rangle}{\partial x_i} \right) + \frac{\partial}{\partial x_q} \left\langle -\rho V_q^{'} V_i^{'} V_j^{'} \right\rangle = 0.
\end{aligned} \tag{5.70}$$

Since these processes vanish in homogeneous turbulence, they are called *inhomogeneous processes*. Similarly, in the transport equation of k, the advection process on the LHS of (5.56), and the processes marked as $IV^{'}$ and $V^{'}$ on the right-hand side of the equation are called *inhomogeneous processes of the transport equation of k*.

5.7 The RSTE and the Turbulence Closure Problem

Earlier in Sect. 5.3, we discussed that addressing the turbulence closure problem of the mean Navier-Stokes equation set (5.10) requires the inclusion of additional equations describing the variation of the Reynolds stress tensor in the turbulent flow field. Indeed, at the first glance, the derived transport equation of the Reynolds stress tensor (5.43) appears to provide six additional scalar equations. These equations may be added to the mean Navier-Stokes equation set (5.10), making the total number of equations ten ($10 = 1 + 3 + 6$). *Does this help us achieve mathematical closure?*

We realize that the transport equation of the Reynolds stress tensor (5.49) has itself introduced several new unknowns. These new unknowns are (i) the six independent scalar components of the $\underline{\epsilon}$ tensor, (ii) the five independent scalar components of the $\underline{\Pi}$ tensor, the three scalar components of the $\left\langle p^{'} \underline{V}^{'} \right\rangle$ vector and the ten independent scalar components of the third-order tensor $\left\langle \underline{V}^{'} \underline{V}^{'} \underline{V}^{'} \right\rangle$. We call these additional 24 scalar quantities as the *tertiary unknowns*. Thus, the equation set comprised of (5.10) and (5.43) now has 10 scalar equations but 34 scalar unknowns. These unknowns are listed as:

$$04 \text{ Primary Unknowns: } \langle V_i \rangle \text{ and } \langle p \rangle$$

$$06 \text{ Secondary unknowns: } \left\langle -\rho V_i^{'} V_j^{'} \right\rangle$$

$$24 \text{ Tertiary Unknowns: } \epsilon_{ij} \text{ , } \Pi_{ij} \text{ , } \left\langle p^{'} V_i^{'} \right\rangle, \text{ and } \left\langle V_i^{'} V_j^{'} V_k^{'} \right\rangle$$

Evidently, even with the inclusion of the exact transport equation of the Reynolds stress tensor (5.49), the turbulence closure problem persists. In fact, any further attempt to derive the exact governing equations for the tertiary unknowns would invariably introduce newer unknowns. Thus, to address the turbulence closure problem, we must adopt an alternate approach. Instead of deriving further exact governing equations for the new unknowns, we should approximately *model* them in terms of the flow variables for which governing equations have already been included.

Historically, several such approximate models have been proposed. Many of these closure models are *phenomenological* in nature. They have been proposed based on our experimental or DNS-based observations of various quantities of interest in some very simple flow fields such as a turbulent boundary layer, a homogeneous shear flow, or a decaying turbulent flow. Despite their simplicity, an experimental or DNS database of such flow fields does provide very useful information with which we can possibly develop deeper insights into the variation of Reynolds stresses and their relationship with the primary flow variables. Such accumulated information and insights are often leveraged while attempting to develop new turbulence closure models. With such motivation, in the next two chapters, we examine two such simple flow fields: (i) turbulent boundary layer (Chap. 6), and (ii) decaying turbulence (Chap. 7). Subsequently, in Chap. 8, where we review a couple of popular turbulence closure models, we will further highlight how some specific observations from such flow fields have been employed to optimize the performance of these models.

Turbulence Near a Solid Wall

6

In this chapter, we examine a turbulent boundary layer that develops over a thin and wide flat plate (Fig. 6.1). The flow field in the far-upstream conditions is $V_0\hat{e}_1$, where V_o is a constant. The plate is fixed to the reference frame in context (the ground frame). We employ a frame-fixed Cartesian coordinate system $Ox_1(\hat{e}_1)x_2(\hat{e}_2)x_3(\hat{e}_3)$. The origin of the coordinate system is fixed at a location such that the boundary layer is turbulent at all locations $x_1 > 0$.

The width of the plate in $\pm\hat{e}_3$ is very large, such that the influence of the plate edges is negligible in the zone of interest. Like any other turbulent flows, the velocity field $\underline{V}(\underline{X}, t)$ and the pressure field $p(\underline{X}, t)$ are unsteady and three-dimensional (3D, which means all flow variables vary with all three spatial coordinates, x_1, x_2 and x_3). Further, generally, the velocity field has three components (3C). The flat plate imposes, like in a laminar boundary layer, the no-slip and no-penetration boundary condition on the velocity field at $x_2 = 0$. Our intent in this chapter is to examine the observed behavior of expectations of various variables of interest within the turbulent boundary layer in the DNS database. Even though the boundary conditions of this flow field are very simple, still the flow field offers an opportunity for us to understand how the presence of a solid wall influences the expectations of various flow variables of interest. Further, this study provides some cues which have helped in the development of turbulence closure models.

6.1 Observed Behavior of the Mean Flow Field

Like any turbulent flow field, the instantaneous velocity and pressure fields are unsteady and three-dimensional, and the velocity field is inherently 3C.

$$\frac{\partial \underline{V}}{\partial t} \neq 0 \text{ and } \frac{\partial p}{\partial t} \neq 0 \tag{6.1}$$

S. S. Sinha, *Fundamentals of Turbulence and Its Modeling*,
https://doi.org/10.1007/978-3-031-94016-3_6

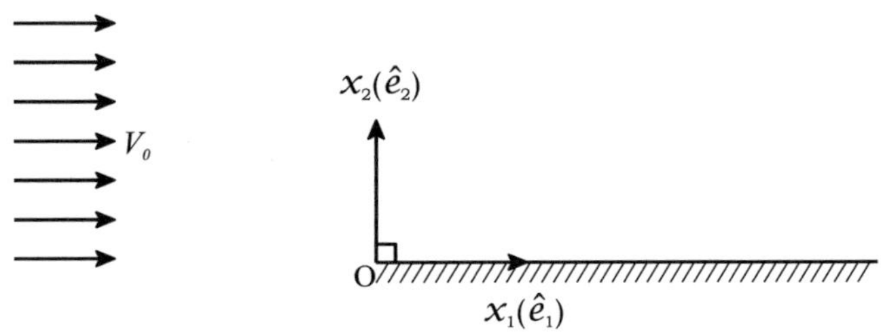

Fig. 6.1 A flat plate placed in an otherwise uniform stream

$$\frac{\partial \underline{V}}{\partial x_i} \neq 0 \text{ and } \frac{\partial p}{\partial x_i} \neq 0 \tag{6.2}$$

where $i = 1, 2$ or 3.

However, the turbulent flow field over a thin and wide plate is known to be statistically stationary and statistically homogeneous in the $\pm\hat{e}_3$ direction (Fig. 6.1). Thus, the time derivative and the spatial gradient along the $\hat{e}_3$ direction of all expected values are zero at all locations and at all time instants.

$$\frac{\partial \langle . \rangle}{\partial t} = 0 \tag{6.3}$$

and

$$\frac{\partial \langle . \rangle}{\partial x_3} = 0 \tag{6.4}$$

Further, the velocity field in this flow is known to be statistically two componential with

$$\langle V_3 \rangle = 0 \Rightarrow \langle \underline{V} \rangle = \langle V_1 \rangle \hat{e}_1 + \langle V_2 \rangle \hat{e}_2 \tag{6.5}$$

Like any other turbulent flow field, the fluctuating components of all flow variables are unsteady and 3D.

$$\frac{\partial \underline{V}'}{\partial t} \neq 0 \text{ and } \frac{\partial p'}{\partial t} \neq 0 \tag{6.6}$$

$$\frac{\partial \underline{V}'}{\partial x_i} \neq 0 \text{ and } \frac{\partial p'}{\partial x_i} \neq 0 \tag{6.7}$$

Further, like any other turbulent flow field, the fluctuating velocity component is 3C (In general, V_1', V_2', and V_3' are all non-zero.)

The local boundary layer thickness $\delta(x_1)$ of a turbulent boundary layer is defined to be the wall-normal distance (x_2) where

$$\langle V_1(x_1, x_2 = \delta(x_1)) \rangle = 0.99 V_o. \tag{6.8}$$

In Fig. 6.2 we present the mean values of various statistics computed using the DNS database of Schlattr and Örlu [3]. The variations in various statistics have been

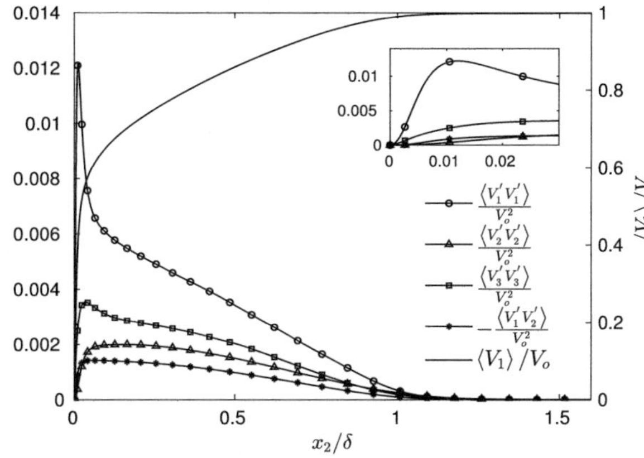

Fig. 6.2 Variation in mean quantities along the wall-normal direction at a stream-wise station where the Reynolds number based on the momentum thickness is 3970. Source of DNS data: [3]. The vertical axis on the right is to be used for $\langle V_1\rangle$. The vertical axis on the left is to be used for the velocity correlation data

shown along the wall-normal distance at a stream-wise (along the plate) station (x_a) where the Reynolds number based on the momentum thickness is 3970.

$$Re_{x_1} = \frac{\rho V_o \theta}{\mu} = 3970 \tag{6.9}$$

where θ represents the local momentum thickness.

The vertical axis on the right-hand side corresponds to the normalized mean stream-wise velocity ($\langle V_1\rangle$), where the normalizing quantity is the far-upstream velocity (represented by the symbol V_o). In the far-upstream locations, the velocity vector is purely stream-wise, and the flow field is laminar, with no fluctuations present in any flow variable. The vertical axis, on the left-hand side, corresponds to the other quantities: $\left\langle V_1^{'}V_1^{'}\right\rangle$, $\left\langle V_2^{'}V_2^{'}\right\rangle$, $\left\langle V_3^{'}V_3^{'}\right\rangle$ and $\left\langle V_1^{'}V_2^{'}\right\rangle$. These quantities have been normalized by V_o^2. On the other hand, the horizontal axis represents the normalized wall-normal distance from the plate, with the normalizing quantity being the local boundary layer thickness ($\delta(x_a)$). There are several pertinent observations.

1. At $x_2 = 0$, $\langle V_1\rangle = 0$ due to the no-slip condition. As $x_2 \to \delta(x_1)$ increases, $\langle V_1\rangle / V_o$ monotonically increases and tends to reach its asymptotic value of unity as $x_2 \to \delta(x_a)$.
2. The slope of the $\langle V_1\rangle$ curve (or $\frac{\partial\langle V_1\rangle}{\partial x_2}$) is observed to be the highest at the wall ($x_2 = 0$) and then subsequently decays monotonically to reach zero as $x_2 \to \delta(x_a)$.
3. In contrast to the variation in the mean stream-wise velocity, the variations in $\left\langle V_1^{'}V_1^{'}\right\rangle$, $\left\langle V_2^{'}V_2^{'}\right\rangle$, $\left\langle V_3^{'}V_3^{'}\right\rangle$ and $\left\langle V_1^{'}V_2^{'}\right\rangle$ are non-monotonic (see the inset zoomed-in view in Fig. 6.2). The no-slip and the no-penetration conditions do ensure that at $x_2 = 0$, all these quantities are zero. However, as one moves away from the wall, there is a sharp increase observed in all these quantities. As the wall-normal distance further increases, the quantities tend to decrease and eventually vanish near the edge of the boundary layer.

4. Close to the wall, among the three components $\left\langle V_1^{'} V_1^{'} \right\rangle$, $\left\langle V_2^{'} V_2^{'} \right\rangle$ and $\left\langle V_3^{'} V_3^{'} \right\rangle$, $\left\langle V_1^{'} V_1^{'} \right\rangle$ seems to have the highest magnitude and $\left\langle V_2^{'} V_2^{'} \right\rangle$ seems to have the smallest magnitude. However, as one approaches the edge of the boundary layer, the disparity in the magnitudes of $\left\langle V_1^{'} V_1^{'} \right\rangle$, $\left\langle V_2^{'} V_2^{'} \right\rangle$ and $\left\langle V_3^{'} V_3^{'} \right\rangle$ tends to disappear and at locations $x_2/\delta(x_a) > 0.9$, we observe

$$\left\langle V_1^{'} V_1^{'} \right\rangle \approx \left\langle V_2^{'} V_2^{'} \right\rangle \approx \left\langle V_3^{'} V_3^{'} \right\rangle \tag{6.10}$$

5. Even though at locations close to the wall, $\left\langle V_1^{'} V_1^{'} \right\rangle$, $\left\langle V_2^{'} V_2^{'} \right\rangle$ and $\left\langle V_3^{'} V_3^{'} \right\rangle$ are not equal, the order of magnitude of all three quantities is still deemed to be the same.
6. The quantity $\left\langle V_1^{'} V_2^{'} \right\rangle$, too, shows non-monotonic variation with the wall-normal distance. It is zero at the wall (at $x_2 = 0$) due to the no-slip and no-penetration conditions. As x_2 increases, it first increases, reaches a peak value, and then subsequently tends to vanish as one approaches the boundary layer edge. Further, $\left\langle V_1^{'} V_2^{'} \right\rangle < 0$ at all locations in the boundary layer.

6.2 Simplified Governing Equations of the Mean Flow Field

To further understand the observations gathered from Fig. 6.2, we refer again to the mean Navier-Stokes equations set (5.2 and 5.8). These equations in their current forms represent the governing equations of a general constant-density turbulent flow field of a Newtonian fluid. For the flow over the flat plate as described in Fig. 6.1, however, this equation set can be subsequently simplified. The first set of simplification invokes properties (6.3), (6.4) and (6.5). Employing (6.5), in the mean continuity equation (5.2), $\frac{\partial \langle V_3 \rangle}{\partial x_3}$ vanishes, leading to the following simplified form of the mean continuity equation

$$\frac{\partial \langle V_1 \rangle}{\partial x_1} + \frac{\partial \langle V_2 \rangle}{\partial x_2} = 0 \tag{6.11}$$

Next, we examine the mean momentum equation along the $\hat{e}_1$ direction

$$\begin{aligned} &\rho \frac{\partial \langle V_1 \rangle}{\partial t} + \rho \langle V_1 \rangle \frac{\partial \langle V_1 \rangle}{\partial x_1} + \rho \langle V_2 \rangle \frac{\partial \langle V_1 \rangle}{\partial x_2} + \rho \langle V_3 \rangle \frac{\partial \langle V_1 \rangle}{\partial x_3} \\ &= -\frac{\partial \langle p \rangle}{\partial x_1} + \mu \frac{\partial^2 \langle V_1 \rangle}{\partial x_1 \partial x_1} + \mu \frac{\partial^2 \langle V_1 \rangle}{\partial x_2 \partial x_2} + \mu \frac{\partial^2 \langle V_1 \rangle}{\partial x_3 \partial x_3} \\ &\quad - \frac{\partial \left\langle \rho V_1^{'} V_1^{'} \right\rangle}{\partial x_1} - \frac{\partial \left\langle \rho V_2^{'} V_1^{'} \right\rangle}{\partial x_2} - \frac{\partial \left\langle \rho V_3^{'} V_1^{'} \right\rangle}{\partial x_3} \end{aligned} \tag{6.12}$$

Using (6.3), (6.4) and (6.5), (6.12) simplifies to

$$\begin{aligned}
&\rho \langle V_1 \rangle \frac{\partial \langle V_1 \rangle}{\partial x_1} + \rho \langle V_2 \rangle \frac{\partial \langle V_1 \rangle}{\partial x_2} \\
&= -\frac{\partial \langle p \rangle}{\partial x_1} + \mu \frac{\partial^2 \langle V_1 \rangle}{\partial x_1 \partial x_1} + \mu \frac{\partial^2 \langle V_1 \rangle}{\partial x_2 \partial x_2} \\
&- \frac{\partial \left\langle \rho V_1' V_1' \right\rangle}{\partial x_1} - \frac{\partial \left\langle \rho V_2' V_1' \right\rangle}{\partial x_2}
\end{aligned} \tag{6.13}$$

Next, we examine the mean momentum equation along the $\hat{e}_2$ direction

$$\begin{aligned}
&\rho \frac{\partial \langle V_2 \rangle}{\partial t} + \rho \langle V_1 \rangle \frac{\partial \langle V_2 \rangle}{\partial x_1} + \rho \langle V_2 \rangle \frac{\partial \langle V_2 \rangle}{\partial x_2} + \rho \langle V_3 \rangle \frac{\partial \langle V_2 \rangle}{\partial x_3} \\
&= -\frac{\partial \langle p \rangle}{\partial x_2} + \mu \frac{\partial^2 \langle V_2 \rangle}{\partial x_1 \partial x_1} + \mu \frac{\partial^2 \langle V_2 \rangle}{\partial x_2 \partial x_2} + \mu \frac{\partial^2 \langle V_2 \rangle}{\partial x_3 \partial x_3} \\
&- \frac{\partial \left\langle \rho V_1' V_2' \right\rangle}{\partial x_1} - \frac{\partial \left\langle \rho V_2' V_2' \right\rangle}{\partial x_2} - \frac{\partial \left\langle \rho V_3' V_2' \right\rangle}{\partial x_3}
\end{aligned} \tag{6.14}$$

Using (6.3), (6.4) and (6.5), (6.14) simplifies to

$$\begin{aligned}
&\rho \langle V_1 \rangle \frac{\partial \langle V_2 \rangle}{\partial x_1} + \rho \langle V_2 \rangle \frac{\partial \langle V_2 \rangle}{\partial x_2} \\
&= -\frac{\partial \langle p \rangle}{\partial x_2} + \mu \frac{\partial^2 \langle V_2 \rangle}{\partial x_1 \partial x_1} + \mu \frac{\partial^2 \langle V_2 \rangle}{\partial x_2 \partial x_2} \\
&- \frac{\partial \left\langle \rho V_1' V_2' \right\rangle}{\partial x_1} - \frac{\partial \left\langle \rho V_2' V_2' \right\rangle}{\partial x_2}
\end{aligned} \tag{6.15}$$

Next, we examine the mean momentum equation along the $\hat{e}_3$ direction

$$\begin{aligned}
&\rho \frac{\partial \langle V_3 \rangle}{\partial t} + \rho \langle V_1 \rangle \frac{\partial \langle V_3 \rangle}{\partial x_1} + \rho \langle V_2 \rangle \frac{\partial \langle V_3 \rangle}{\partial x_2} + \rho \langle V_3 \rangle \frac{\partial \langle V_3 \rangle}{\partial x_3} \\
&= -\frac{\partial \langle p \rangle}{\partial x_3} + \mu \frac{\partial^2 \langle V_3 \rangle}{\partial x_1 \partial x_1} + \mu \frac{\partial^2 \langle V_3 \rangle}{\partial x_2 \partial x_2} + \mu \frac{\partial^2 \langle V_3 \rangle}{\partial x_3 \partial x_3} \\
&- \frac{\partial \left\langle \rho V_1' V_3' \right\rangle}{\partial x_1} - \frac{\partial \left\langle \rho V_2' V_3' \right\rangle}{\partial x_2} - \frac{\partial \left\langle \rho V_3' V_3' \right\rangle}{\partial x_3}
\end{aligned} \tag{6.16}$$

Using (6.3), (6.4) and (6.5), (6.16) simplifies to

$$0 = -\frac{\partial \left\langle \rho V_1' V_3' \right\rangle}{\partial x_1} - \frac{\partial \left\langle \rho V_2' V_3' \right\rangle}{\partial x_2} \tag{6.17}$$

The simplified form of $\langle V_3 \rangle$ involves merely the spatial gradients of the two specific components of the Reynolds stress tensor: $\left\langle -\rho V_1^{'} V_3^{'} \right\rangle$ and $\left\langle -\rho V_2^{'} V_3^{'} \right\rangle$. Since $\langle V_3 \rangle$ is already known to be zero for this flow field, we do not have any immediate motivation to further analyze (6.17).

6.3 Order-of-Magnitude Analysis of the Governing Equations

Equations (6.11), (6.13), and (6.15) can be further simplified by performing an order-of-magnitude (OM) analysis of these equations. This analysis is similar to the one performed for a laminar boundary layer [1]. However, care must be taken to use appropriate characteristic values for the mean velocity components and the relevant Reynolds stress components.

We say that ϕ_C is the characteristic value of the mean of the variable $\phi(x_1, x_2, x_3, t)$ over a domain of interest, if

$$\mathcal{O}\left[\frac{\langle \phi \rangle}{\phi_C}\right] = 10^0 = 1 \tag{6.18}$$

over most of the domains of interest. Using ϕ_C, we now define a normalized version (denoted with a superscript $*$) of the mean variable

$$\langle \phi \rangle^* = \frac{\langle \phi \rangle}{\phi_C} \tag{6.19}$$

Equation (6.18) implies that if ϕ_C has been aptly chosen, then

$$\mathcal{O}\left[\langle \phi \rangle^*\right] = 1 \tag{6.20}$$

over most of the domains of interest.

In the context of the governing equations of the flow over a flat plate, if we appropriately choose two numbers V_{1C} and V_{2C} as the characteristic values of the variables $\langle V_1 \rangle$ and $\langle V_2 \rangle$, then we can define the normalized version of these variables as

$$\langle V_1 \rangle^* = \frac{\langle V_1 \rangle}{V_{1C}} \text{ and } \langle V_2 \rangle^* = \frac{\langle V_2 \rangle}{V_{2C}} \tag{6.21}$$

such that

$$\mathcal{O}\left[\langle V_1 \rangle^*\right] = 1 \text{ and } \mathcal{O}\left[\langle V_2 \rangle^*\right] = 1 \tag{6.22}$$

over most of the domains of interest. Here, $\langle V_1 \rangle^*$ and $\langle V_2 \rangle^*$ are still variables of x_1 and x_2, but they are not dependent on time (the flow being statistically stationary) or x_3 (the flow being statistically homogeneous along the $\hat{e}_3$ direction). For the boundary layer flow, we choose

$$V_{1C} = V_o \tag{6.23}$$

To choose appropriate characteristic values of the Reynolds stress components, we refer back to our observations in Fig. 6.2. Specifically, our observation included in (6.10) suggests that $\rho a^2 V_o^2$ could be an apt choice of the characteristic value for the normal Reynolds stress components, where a is an appropriately chosen positive constant, such that,

$$\mathcal{O}[a] = 1 \text{ and } a < 1 \tag{6.24}$$

Thus, the corresponding normalized forms of these stress components are defined

$$\begin{aligned}
\left\langle \rho V_1' V_1' \right\rangle^* &= \frac{\left\langle \rho V_1' V_1' \right\rangle}{\rho a^2 V_{1C}^2} \\
\left\langle \rho V_2' V_2' \right\rangle^* &= \frac{\left\langle \rho V_2' V_2' \right\rangle}{\rho a^2 V_{1C}^2} \\
\left\langle \rho V_3' V_3' \right\rangle^* &= \frac{\left\langle \rho V_3' V_3' \right\rangle}{\rho a^2 V_{1C}^2}
\end{aligned} \tag{6.25}$$

such that

$$\mathcal{O}\left\langle \rho V_1' V_1' \right\rangle^* = 1,\ \mathcal{O}\left\langle \rho V_2' V_2' \right\rangle^* = 1 \text{ and } \mathcal{O}\left\langle \rho V_3' V_3' \right\rangle^* = 1 \tag{6.26}$$

Equation (6.25) implies that

$$\mathcal{O}\sqrt{\langle V_1' V_1' \rangle} = \mathcal{O}\sqrt{\langle V_2' V_2' \rangle} = \mathcal{O}\sqrt{\langle V_3' V_3' \rangle} = \mathcal{O}\left[a V_o\right] \tag{6.27}$$

Based on these observations, we conjecture

$$\mathcal{O}\left[\left\langle V_1' V_2' \right\rangle\right] \leq \mathcal{O}\left[a^2 V_o^2\right] \tag{6.28}$$

Thus, we normalize the shear stress Reynolds stress components, as well, using the same characteristic value which has been used for normalizing the normal stress components

$$\begin{aligned}
\left\langle \rho V_1' V_2' \right\rangle^* &= \frac{\left\langle \rho V_1' V_2' \right\rangle}{\rho a^2 V_{1C}^2} \\
\left\langle \rho V_2' V_3' \right\rangle^* &= \frac{\left\langle \rho V_2' V_3' \right\rangle}{\rho a^2 V_{1C}^2} \\
\left\langle \rho V_3' V_1' \right\rangle^* &= \frac{\left\langle \rho V_3' V_1' \right\rangle}{\rho a^2 V_{1C}^2}
\end{aligned} \tag{6.29}$$

such that

$$\mathcal{O}\left\langle \rho V_1' V_2' \right\rangle^* \leq 1, \mathcal{O}\left\langle \rho V_2' V_3' \right\rangle^* \leq 1 \text{ and } \mathcal{O}\left\langle \rho V_3' V_1' \right\rangle^* \leq 1 \tag{6.30}$$

Further, we define *characteristic length-scales*, L_{1C} and L_{2C}, for the mean flow inside the boundary layer such that with these length-scales, the gradients of various scalar variables along the $\hat{e}_1$ and $\hat{e}_2$ directions are to normalized as

$$\begin{aligned}
&\left(\frac{\partial \langle V_1 \rangle}{\partial x_1}\right)^* = \left(\frac{\partial \langle V_1 \rangle}{\partial x_1}\right)\frac{L_{1C}}{V_{1C}}, \left(\frac{\partial \langle V_1 \rangle}{\partial x_2}\right)^* = \left(\frac{\partial \langle V_1 \rangle}{\partial x_2}\right)\frac{L_{2C}}{V_{1C}}, \\
&\left(\frac{\partial \langle V_2 \rangle}{\partial x_1}\right)^* = \left(\frac{\partial \langle V_2 \rangle}{\partial x_1}\right)\frac{L_{1C}}{V_{2C}}, \left(\frac{\partial \langle V_2 \rangle}{\partial x_2}\right)^* = \left(\frac{\partial \langle V_2 \rangle}{\partial x_2}\right)\frac{L_{2C}}{V_{2C}} \\
&\left(\frac{\partial \left\langle \rho V_i' V_j' \right\rangle}{\partial x_1}\right)^* = \left(\frac{\partial \left\langle \rho V_i' V_j' \right\rangle}{\partial x_1}\right)\frac{L_{1C}}{\rho a^2 V_{1C}^2} \\
&\left(\frac{\partial \left\langle \rho V_i' V_j' \right\rangle}{\partial x_2}\right)^* = \left(\frac{\partial \left\langle \rho V_i' V_j' \right\rangle}{\partial x_2}\right)\frac{L_{2C}}{\rho a^2 V_{1C}^2}
\end{aligned} \tag{6.31}$$

such that

$$\begin{aligned}
&\mathcal{O}\left[\left(\frac{\partial \langle V_1 \rangle}{\partial x_1}\right)^*\right] = 1, \mathcal{O}\left[\left(\frac{\partial \langle V_1 \rangle}{\partial x_2}\right)^*\right] = 1, \\
&\mathcal{O}\left[\left(\frac{\partial \langle V_2 \rangle}{\partial x_1}\right)^*\right] = 1, \mathcal{O}\left[\left(\frac{\partial \langle V_2 \rangle}{\partial x_2}\right)^*\right] = 1 \\
&\mathcal{O}\left[\left(\frac{\partial \left\langle V_1' V_1' \right\rangle}{\partial x_1}\right)^*\right] = 1, \mathcal{O}\left[\left(\frac{\partial \left\langle V_1' V_1' \right\rangle}{\partial x_2}\right)^*\right] = 1 \\
&\mathcal{O}\left[\left(\frac{\partial \left\langle V_2' V_2' \right\rangle}{\partial x_1}\right)^*\right] = 1, \mathcal{O}\left[\left(\frac{\partial \left\langle V_2' V_2' \right\rangle}{\partial x_2}\right)^*\right] = 1 \\
&\mathcal{O}\left[\left(\frac{\partial \left\langle V_3' V_3' \right\rangle}{\partial x_1}\right)^*\right] = 1, \mathcal{O}\left[\left(\frac{\partial \left\langle V_3' V_3' \right\rangle}{\partial x_2}\right)^*\right] = 1
\end{aligned} \tag{6.32}$$

and

$$\mathcal{O}\left[\left(\frac{\partial\left\langle V_1'V_2'\right\rangle}{\partial x_1}\right)^*\right]\le 1,\ \mathcal{O}\left[\left(\frac{\partial\left\langle V_1'V_2'\right\rangle}{\partial x_2}\right)^*\right]\le 1$$
$$\mathcal{O}\left[\left(\frac{\partial\left\langle V_2'V_3'\right\rangle}{\partial x_1}\right)^*\right]\le 1,\ \mathcal{O}\left[\left(\frac{\partial\left\langle V_2'V_3'\right\rangle}{\partial x_2}\right)^*\right]\le 1$$
$$\mathcal{O}\left[\left(\frac{\partial\left\langle V_3'V_1'\right\rangle}{\partial x_1}\right)^*\right]\le 1,\ \mathcal{O}\left[\left(\frac{\partial\left\langle V_3'V_1'\right\rangle}{\partial x_2}\right)^*\right]\le 1 \tag{6.33}$$

Further, using the characteristic values of the velocity components (V_{1C}, V_{2C}) and the characteristic length-scales (L_{1C}, L_{2C}), we define the normalized versions of the second derivatives of the velocity components

$$\left(\frac{\partial^2\langle V_1\rangle}{\partial x_1\partial x_1}\right)^*=\left(\frac{\partial^2\langle V_1\rangle}{\partial x_1\partial x_1}\right)\frac{L_{1C}^2}{V_{1C}},\ \left(\frac{\partial^2\langle V_1\rangle}{\partial x_2\partial x_2}\right)^*=\left(\frac{\partial^2\langle V_1\rangle}{\partial x_2\partial x_2}\right)\frac{L_{2C}^2}{V_{1C}},$$
$$\left(\frac{\partial^2\langle V_2\rangle}{\partial x_1\partial x_1}\right)^*=\left(\frac{\partial^2\langle V_2\rangle}{\partial x_1\partial x_1}\right)\frac{L_{1C}^2}{V_{2C}},\ \left(\frac{\partial^2\langle V_2\rangle}{\partial x_2\partial x_2}\right)^*=\left(\frac{\partial^2\langle V_2\rangle}{\partial x_2\partial x_2}\right)\frac{L_{2C}^2}{V_{2C}}, \tag{6.34}$$

such that

$$\mathcal{O}\left[\left(\frac{\partial^2\langle V_1\rangle}{\partial x_1\partial x_1}\right)^*\right]=1,\ \mathcal{O}\left[\left(\frac{\partial^2\langle V_1\rangle}{\partial x_2\partial x_2}\right)^*\right]=1,$$
$$\mathcal{O}\left[\left(\frac{\partial^2\langle V_2\rangle}{\partial x_1\partial x_1}\right)^*\right]=1,\ \mathcal{O}\left[\left(\frac{\partial^2\langle V_2\rangle}{\partial x_2\partial x_2}\right)^*\right]=1 \tag{6.35}$$

To perform the OM analysis of (6.11), (6.13), and (6.15), we implement the following three steps separately for each of these three equations.

1. We substitute the dimensional variables and their derivatives by their non-dimensional counterparts using (6.21), (6.31), and (6.34).
2. Using the anticipation (which is based on the appropriately chosen characteristic values of the velocity components V_{1C} and V_{2C} and the length-scales L_{1C} and L_{2C}) that all normalized versions of the variables would have their orders of magnitude unity over most of the domain of interest, we identify those terms (if any) which have smaller orders of magnitude than the other additive terms of the equation. These identified terms can then be discarded from the equation.
3. We substitute the normalized versions of the variables by their dimensional counterparts using (6.21, (6.31), and (6.34) and arrive at a simplified version of the governing equation.

The order-of-magnitude analysis of the governing equations is performed at an arbitrary location (x_1, x_2, x_3) and at an arbitrary time (t) inside the turbulent boundary layer region. In the neighborhood of such a location, we deem the following choices of the length-scales to be appropriate

$$\begin{aligned}\mathcal{O}[L_{1C}] &= \mathcal{O}[x_1], \\ \mathcal{O}[L_{2C}] &= \mathcal{O}[\delta(x_1)]\end{aligned} \tag{6.36}$$

Accordingly, we choose,

$$L_{1C} = x_1 \text{ and } L_{2C} = \delta(x_1) \tag{6.37}$$

Further, we deem V_o to be an appropriate choice of V_{1C}. However, at this point, we do not have any cues available to choose V_{2C}. Nonetheless, we do initiate our OM analysis and move on to explore if any further cues emerge to choose V_{2C} appropriately.

6.3.1 Continuity Equation

Substituting the raw variables by their normalized counterparts in the continuity equation (6.11) results in the following form of the equation

$$\left(\frac{\partial \langle V_1\rangle}{\partial x_1}\right)^* \frac{V_{1C}}{L_{1C}} + \left(\frac{\partial \langle V_2\rangle}{\partial x_2}\right)^* \frac{V_{2C}}{L_{2C}} = 0 \tag{6.38}$$

Rearranging various factors, we arrive at

$$\left(\frac{\partial \langle V_1\rangle}{\partial x_1}\right)^* + \left(\frac{\partial \langle V_2\rangle}{\partial x_2}\right)^* \frac{V_{2C}}{V_{1C}} \frac{L_{1C}}{L_{2C}} = 0 \tag{6.39}$$

The first term on the LHS in (6.39) has its order-of-magnitude unity. Further, since the RHS of (6.39) is zero and there are only two terms on the LHS, the order of magnitude of the second term on the LHS must also be unity.

$$\mathcal{O}\left[\left(\frac{\partial \langle V_2\rangle}{\partial x_2}\right)^* \frac{V_{2C}}{V_{1C}} \frac{L_{1C}}{L_{2C}}\right] = 1 \tag{6.40}$$

Since the order of magnitude of the $\left(\frac{\partial\langle V_2\rangle}{\partial x_2}\right)^*$ itself is unity, (6.40) leads to the following conclusion

$$\mathcal{O}\left[\frac{V_{2C}}{V_{1C}} \frac{L_{1C}}{L_{2C}}\right] = 1 \tag{6.41}$$

This, in turn, gives us a cue to choose V_{2C}.

$$V_{2C} = V_{1C}\frac{L_{2C}}{L_{1C}} \tag{6.42}$$

Thus, even though the order-of-magnitude analysis of the continuity equation does not provide any justification to simplify the equation any further, it has provided a justification based on which we now have a characteristic value of the variable $\langle V_2 \rangle$ in terms V_{1C}, L_{2C} and L_{1C}.

6.3.2 The $\langle V_1 \rangle$ Equation

We now substitute the raw variables by their normalized counterparts in the governing equation of $\langle V_1 \rangle$ (6.13)

$$\begin{aligned}
&\left[\langle V_1\rangle^* V_{1C}\left(\frac{\partial \langle V_1\rangle}{\partial x_1}\right)^*\frac{V_{1C}}{L_{1C}} + \langle V_2\rangle^* V_{2C}\left(\frac{\partial \langle V_1\rangle}{\partial x_2}\right)^*\frac{V_{1C}}{L_{2C}}\right] = \\
&\qquad -\frac{1}{\rho}\frac{\partial \langle p\rangle}{\partial x_1} + \nu\left(\frac{\partial^2 \langle V_1\rangle}{\partial x_1 \partial x_1}\right)^*\frac{V_{1C}}{L_{1C}^2} + \nu\left(\frac{\partial^2 \langle V_1\rangle}{\partial x_2 \partial x_2}\right)^*\frac{V_{1C}}{L_{2C}^2} \\
&-\frac{1}{\rho}\left(\frac{\partial\left\langle \rho V_1' V_1'\right\rangle}{\partial x_1}\right)^*\frac{\rho a^2 V_{1C}^2}{L_{1C}} - \frac{1}{\rho}\left(\frac{\partial\left\langle \rho V_1' V_2'\right\rangle}{\partial x_2}\right)^*\frac{\rho a^2 V_{1C}^2}{L_{2C}}
\end{aligned} \tag{6.43}$$

Using (6.42) to substitute V_{2C} in terms of V_{1C}, L_{2C} and L_{1C} in (6.43) leads to the following form of the equation.

$$\begin{aligned}
&\frac{V_{1C}^2}{L_{1C}}\left[\langle V_1\rangle^*\left(\frac{\partial \langle V_1\rangle}{\partial x_1}\right)^* + \langle V_2\rangle^*\left(\frac{\partial \langle V_1\rangle}{\partial x_2}\right)^*\right] = \\
&\qquad -\frac{1}{\rho}\frac{\partial \langle p\rangle}{\partial x_1} + \left(\frac{\partial^2 \langle V_1\rangle}{\partial x_1 \partial x_1}\right)^*\frac{\nu V_{1C}}{L_{1C}^2} + \left(\frac{\partial^2 \langle V_1\rangle}{\partial x_2 \partial x_2}\right)^*\frac{\nu V_{1C}}{L_{2C}^2} \\
&-\frac{1}{\rho}\left(\frac{\partial\left\langle \rho V_1' V_1'\right\rangle}{\partial x_1}\right)^*\frac{\rho a^2 V_{1C}^2}{L_{1C}} - \frac{1}{\rho}\left(\frac{\partial\left\langle \rho V_1' V_2'\right\rangle}{\partial x_2}\right)^*\frac{\rho a^2 V_{1C}^2}{L_{2C}}
\end{aligned} \tag{6.44}$$

The LHS of (6.44) is the stream-wise component of the mean acceleration of a fluid particle within the boundary layer. The expression within the square parentheses represents the normalized form of this acceleration. Clearly, the order of magnitude (OM) of this normalized form is unity. Thus, the quantity V_{1C}^2/L_{1C} (the factor outside the square parentheses) can be deemed as the characteristic value of the stream-wise component of the mean stream-wise acceleration, itself. Similarly, on the RHS, the quantities $\nu V_{1C}/L_{1C}^2$ and $\nu V_{1C}/L_{2C}^2$ are the characteristic values of the two viscous force (per unit mass) terms acting on a typical fluid element inside the boundary layer. Thus, we conclude that the OM of two terms involving the Reynolds stress components are $a^2 V_{1C}^2/L_{1C}$ and $a^2 V_{1C}^2/L_{2C}$, respectively.

Dividing (6.44) throughout by the factor $\frac{V_{1C}^2}{L_{1C}}$ and further re-arranging the terms, we arrive at the following form of the equation

$$\begin{aligned}\langle V_1\rangle^* \left(\frac{\partial \langle V_1\rangle}{\partial x_1}\right)^* + \langle V_2\rangle^* \left(\frac{\partial \langle V_1\rangle}{\partial x_2}\right)^* &= \left(\frac{\partial^2 \langle V_1\rangle}{\partial x_1 \partial x_1}\right)^* \frac{\nu}{V_{1C} L_{1C}} \\ &+ \left(\frac{\partial^2 V_1}{\partial x_2 \partial x_2}\right)^* \frac{\nu}{V_{1C} L_{1C}} \left(\frac{L_{1C}}{L_{2C}}\right)^2 - \frac{\partial \langle p\rangle}{\partial x_1} \frac{L_{1C}}{\rho V_{1C}^2} \\ &- \left(\frac{\partial \left\langle \rho V_1' V_1' \right\rangle}{\partial x_1}\right)^* a^2 - \left(\frac{\partial \left\langle \rho V_1' V_2' \right\rangle}{\partial x_2}\right)^* \frac{a^2 L_{1C}}{L_{2C}}\end{aligned} \tag{6.45}$$

The LHS of (6.45) of the equation is the fully normalized form of the mean stream-wise acceleration, with its OM being unity. The RHS of (6.45) represents the sum of the ratios of various forces (per unit mass, along the $\hat{e}_1$ direction) to the characteristic value of the mean stream-wise acceleration of a fluid particle (V_{1C}^2/L_{1C}). If any of these terms on the RHS has its order of magnitude $\ll 1$, then we can conclude that the particular force is not significant in contributing toward the stream-wise acceleration of the fluid particle. Thus, that term can be neglected in the equation.

The first two terms on the RHS of (6.45) are the respective ratios of the two viscous force terms to the characteristic value of mean stream-wise acceleration. Both these terms have a common non-dimensional factor involving ν, V_{1C}, and L_{1C}. We define the inverse of this factor as the *Reynolds number of the governing equation of the stream-wise velocity* component. We represent this quantity by the symbol Re_1

$$Re_1 = \frac{V_{1C} L_{1C}}{\nu} \tag{6.46}$$

Based on this new symbol, the order of magnitude of the first term on the RHS of (6.45) can be expressed as:

$$\mathcal{O}\left[\left(\frac{\partial^2 \langle V_1\rangle}{\partial x_1 \partial x_1}\right)^* \frac{\nu}{V_{1C} L_{1C}}\right] = \mathcal{O}\left[\frac{\nu}{V_{1C} L_{1C}}\right] = \mathcal{O}\left[\frac{1}{Re_1}\right] \tag{6.47}$$

since $\mathcal{O}\left[\left(\frac{\partial^2 \langle V_1\rangle}{\partial x_1 \partial x_1}\right)^*\right] = 1$. The quantity Re_1 is interpreted as the ratio of the characteristic value of the mean stream-wise mean acceleration (V_{1C}^2/L_{1C}) of a typical fluid particle inside the boundary layer to the characteristic value of the mean viscous force (per unit mass), along the $\hat{e}_1$ direction. For a turbulent boundary layer, $Re_1 \gg 1$. Accordingly, the first term on the RHS of (6.45) can be neglected.

The factor Re_1 appears in the denominator of the second viscous term of (6.45), as well. However, unlike the first viscous term, the order of magnitude of the second term is not controlled by Re_1 alone but also by the ratio of two relevant length-scales L_{1C} and L_{2C}.

$$\mathcal{O}\left[\left(\frac{\partial^2 \langle V_1\rangle}{\partial x_2 \partial x_2}\right)^* \frac{\nu}{V_{1C} L_{1C}} \left(\frac{L_{1C}}{L_{2C}}\right)^2\right] = \mathcal{O}\left[\left(\frac{\nu}{V_{1C} L_{1C}}\right)\left(\frac{L_{1C}}{L_{2C}}\right)^2\right],$$
$$= \mathcal{O}\left[\frac{1}{Re_1}\left(\frac{L_{1C}}{L_{2C}}\right)^2\right] \tag{6.48}$$

Since the mean flow inside the boundary layer is indeed affected by viscous forces and further given the fact that at $Re_1 \gg 1$, the first viscous term on the RHS of (6.45) is anyway negligible, the second and the only surviving viscous term must be of significance in the evolution equation of $\langle V_1\rangle$. Thus, we conclude that the order of magnitude in (6.48) must itself be unity (same as the order of magnitude of the term included in (6.45)

$$\mathcal{O}\left[\frac{1}{Re_1}\left(\frac{L_{1C}}{L_{2C}}\right)^2\right] = 1 \tag{6.49}$$

This in turn allows estimating L_{2C} in terms of other known parameters (L_{1C} and Re_1)

$$\mathcal{O}\left[\frac{L_{2C}}{L_{1C}}\right] = \mathcal{O}\left[\frac{1}{\sqrt{Re_1}}\right] \tag{6.50}$$

Combining (6.50) with our anticipations in (6.37) we have

$$\mathcal{O}\left[\frac{\delta(x_1)}{x_1}\right] = \mathcal{O}\left[\frac{1}{\sqrt{Re_1}}\right] \tag{6.51}$$

Equation (6.51) suggests that the boundary layer thickness tends to grow as we move more downstream over the plate. Further, since $Re_1 \gg 1$, (6.51) also suggests that the local boundary layer thickness is a small quantity compared to x_1.

The third term on the RHS of (6.45) represents the ratio of the pressure gradient force (per unit mass, along the $\hat{e}_1$ direction) to the characteristic value of streamwise acceleration. Since, at this point, we do not have any reasonable estimate of the characteristic value or the order of magnitude of the pressure gradient term itself, we simply retain this term in the V_1 equation.

The non-dimensional coefficients in the fourth and the fifth terms in (6.44) are a^2 and $a^2 L_{1C}/L_{2C}$. Using the estimate of (6.51), it is evident that

$$\mathcal{O}\left[a^2 L_{1C}/L_{2C}\right] = \mathcal{O}\left[a^2\sqrt{Re_{x_1}}\right] \gg \mathcal{O}\left[a^2\right] \tag{6.52}$$

when $Re_{x_1} \gg 1$. Thus, we can make the approximation

$$-\left(\frac{\partial\left\langle\rho V_1' V_1'\right\rangle}{\partial x_1}\right)^* a^2 - \left(\frac{\partial\left\langle\rho V_1' V_2'\right\rangle}{\partial x_2}\right)^* \frac{a^2 L_{1C}}{L_{2C}} \approx -\left(\frac{\partial\left\langle\rho V_1' V_2'\right\rangle}{\partial x_2}\right)^* \frac{a^2 L_{1C}}{L_{2C}} \tag{6.53}$$

Since $L_{2C}/L_{1C} \ll 1$ at $Re_{x_1} \gg 1$ (6.50), it is evident that the fifth term in (6.45) is much greater than the fourth term, leaving the former as the only representative of the Reynolds stress tensor in the governing equation of $\langle V_1 \rangle$. Since we do expect the Reynolds stress tensor to have some influence on $\langle V_1 \rangle$ within a turbulent boundary layer after all, we retain this term even in the simplified form of (6.45).

In summary, the order-of-magnitude analysis of the $\langle V_1 \rangle$ equation inside the boundary layer over a flat plate has led to the following conclusions:

1. At $Re_1 \gg 1$, the viscous term arising due to the gradient of the stream-wise velocity component with respect to the stream-wise direction can be neglected.
2. If $Re_1 \gg 1$, the Reynolds stress term with a gradient along the stream-wise direction is negligible compared to other Reynolds stress term which appear with a gradient along the wall-normal direction. These approximations simplify (6.13) to

$$\langle V_1 \rangle \frac{\partial \langle V_1 \rangle}{\partial x_1} + \langle V_2 \rangle \frac{\partial \langle V_1 \rangle}{\partial x_2} \approx -\frac{1}{\rho}\frac{\partial \langle p \rangle}{\partial x_1} + \nu \frac{\partial^2 \langle V_1 \rangle}{\partial x_2 \partial x_2} - \frac{\partial\left\langle\rho V_1' V_2'\right\rangle}{\partial x_2}. \tag{6.54}$$

6.3.3 The $\langle V_2 \rangle$ Equation

We substitute the raw variables by their normalized counterparts in the governing equation of $\langle V_2 \rangle$ (6.15)

$$\begin{aligned}\langle V_1 \rangle^* V_{1C} \left(\frac{\partial \langle V_2 \rangle}{\partial x_1}\right)^* \frac{V_{2C}}{L_{1C}} + \langle V_2 \rangle^* V_{2C} \left(\frac{\partial \langle V_2 \rangle}{\partial x_2}\right)^* \frac{V_{2C}}{L_{2C}} = \\ -\frac{1}{\rho}\frac{\partial \langle p \rangle}{\partial x_2} + \nu \left(\frac{\partial^2 \langle V_2 \rangle}{\partial x_1 \partial x_1}\right)^* \frac{V_{2C}}{L_{1C}^2} + \nu \left(\frac{\partial^2 \langle V_2 \rangle}{\partial x_2 \partial x_2}\right)^* \frac{V_{2C}}{L_{2C}^2} \\ -\frac{1}{\rho}\left(\frac{\partial\left\langle\rho V_1' V_2'\right\rangle}{\partial x_1}\right)^* \frac{\rho a^2 V_{1C}^2}{L_{1C}} - \frac{1}{\rho}\left(\frac{\partial\left\langle\rho V_2' V_2'\right\rangle}{\partial x_2}\right)^* \frac{\rho a^2 V_{1C}^2}{L_{2C}}\end{aligned} \tag{6.55}$$

We use (6.42) to substitute V_{2C} in terms of V_{1C}, L_{2C}, and L_{1C} in (6.55), which leads to the following form of the equation:

$$\frac{V_{1C}^2}{L_{1C}}\left(\frac{L_{2C}}{L_{1C}}\right)\left[\langle V_1\rangle^*\left(\frac{\partial\langle V_2\rangle}{\partial x_1}\right)^* + \langle V_2\rangle^*\left(\frac{\partial V_2}{\partial x_2}\right)^*\right] =$$
$$\nu\left(\frac{\partial^2\langle V_2\rangle}{\partial x_1\partial x_1}\right)^*\frac{V_{1C}}{L_{1C}^2}\left(\frac{L_{2C}}{L_{1C}}\right) + \nu\left(\frac{\partial^2\langle V_2\rangle}{\partial x_2\partial x_2}\right)^*\frac{V_{1C}}{L_{2C}^2}\left(\frac{L_{2C}}{L_{1C}}\right) - \frac{1}{\rho}\frac{\partial\langle p\rangle}{\partial x_2}$$
$$-\frac{1}{\rho}\left(\frac{\partial\left\langle\rho V_1' V_2'\right\rangle}{\partial x_1}\right)^*\frac{\rho a^2 V_{1C}^2}{L_{1C}} - \frac{1}{\rho}\left(\frac{\partial\left\langle\rho V_2' V_2'\right\rangle}{\partial x_2}\right)^*\frac{\rho a^2 V_{1C}^2}{L_{2C}} \tag{6.56}$$

The LHS of (6.56) is the wall-normal component of the acceleration of a typical fluid particle within the boundary layer. The expression within the square parentheses represents the normalized form of this acceleration. Clearly, the OM of this normalized form is unity. Thus, the quantity $\left(V_{1C}^2/L_{1C}\right)\left(L_{2C}/L_{1C}\right)$ (the factor outside the square parentheses) is the characteristic value of the wall-normal acceleration component. Since the characteristic value of stream-wise acceleration is already known to be V_{1C}^2/L_{1C}, the LHS of (6.56) implies that the characteristic value of wall-normal component of the acceleration vector is L_{2C}/L_{1C} times the characteristic value of stream-wise acceleration. The discussion of the previous section has already demonstrated that at $Re_1 \gg 1$, L_{2C}/L_{1C} is quite small (6.51). Thus, the net acceleration vector of a typical fluid particle inside the boundary layer is oriented *almost* in the stream-wise direction, allowing us to neglect the acceleration term in the $\langle V_2\rangle$ equation and approximate the rest of the equation as a mere force balance in the wall-normal direction.

Further, even within the scope of this approximate force balance, we wish to examine if all the forces (the two viscous terms, the pressure gradient terms, and the Reynolds stress terms) are individually significant or not. For this analysis, we first divide (6.56) throughout by V_{1C}^2/L_{1C} and arrive at the following form of the $\langle V_2\rangle$ equation.

$$\left[\langle V_1\rangle^*\left(\frac{\partial\langle V_2\rangle}{\partial x_1}\right)^* + \langle V_2\rangle^*\left(\frac{\partial\langle V_2\rangle}{\partial x_2}\right)^*\right]\left(\frac{L_{2C}}{L_{1C}}\right)$$
$$= \left(\frac{\partial^2\langle V_2\rangle}{\partial x_1\partial x_1}\right)^*\frac{1}{Re_1}\left(\frac{L_{2C}}{L_{1C}}\right) + \left(\frac{\partial^2\langle V_2\rangle}{\partial x_2\partial x_2}\right)^*\frac{1}{Re_1}\left(\frac{L_{1C}}{L_{2C}}\right) - \frac{1}{\rho}\frac{\partial\langle p\rangle}{\partial x_2}\frac{L_{1C}}{V_{1C}^2}$$
$$-\left(\frac{\partial\left\langle V_1' V_2'\right\rangle}{\partial x_1}\right)^* a^2 - \left(\frac{\partial\left\langle\rho V_2' V_2'\right\rangle}{\partial x_2}\right)^*\frac{a^2 L_{1C}}{L_{2C}} \tag{6.57}$$

At $Re_1 \gg 1$, the order of magnitude of the first term on the rhs is,

$$\mathcal{O}\left[\left(\frac{\partial^2\langle V_2\rangle}{\partial x_1\partial x_1}\right)^*\frac{1}{Re_1}\left(\frac{L_{2C}}{L_{1C}}\right)\right] = \mathcal{O}\left[\frac{1}{Re_1}\left(\frac{L_{2C}}{L_{1C}}\right)\right] = \mathcal{O}\left[\frac{1}{Re_1^{3/2}}\right] \tag{6.58}$$

which is much less than unity at $Re_1 \gg 1$. Thus, this represents a force component (per unit mass), which is much smaller than the stream-wise acceleration of a typical fluid particle inside the boundary layer.

Further, the order of magnitude of the second term on the RHS of (6.57) is estimated as,

$$\mathcal{O}\left[\left(\frac{\partial^2 \langle V_2\rangle}{\partial x_2 \partial x_2}\right)^* \frac{1}{Re_1}\left(\frac{L_{1C}}{L_{2C}}\right)\right] = \mathcal{O}\left[\frac{1}{Re_1}\left(Re_1^{1/2}\right)\right] = \mathcal{O}\left[\frac{1}{Re_1^{1/2}}\right] \tag{6.59}$$

which, again, is much smaller than unity at high Re_1. Thus, this too represents a force (per unit mass), which is much smaller than the stream-wise acceleration of a typical fluid particle inside the boundary layer. The third term on the RHS of (6.57) is the ratio of the net pressure force (per unit mass) acting on a particle in the wall-normal direction to the characteristic value of stream-wise acceleration. Since, at this point, we do not have any reasonable estimate of the order of magnitude/ characteristic value of the pressure gradient term itself, we simply retain this term in the equation.

The non-dimensional coefficients in the fourth and the fifth terms on the RHS of (6.57) are a^2 and $a^2 L_{1C}/L_{2C}$. Using the estimate in (6.51), it is evident that

$$\mathcal{O}\left[a^2 L_{1C}/L_{2C}\right] = \mathcal{O}\left[a^2\sqrt{Re_{x_1}}\right] \gg \mathcal{O}\left[a^2\right] \tag{6.60}$$

when $Re_1 \gg 1$. Thus, we can make the approximation

$$-\left(\frac{\partial\left\langle\rho V_1' V_1'\right\rangle}{\partial x_1}\right)^* a^2 - \left(\frac{\partial\left\langle\rho V_1' V_2'\right\rangle}{\partial x_2}\right)^* \frac{a^2 L_{1C}}{L_{2C}} \approx - \left(\frac{\partial\left\langle\rho V_1' V_2'\right\rangle}{\partial x_2}\right)^* \frac{a^2 L_{1C}}{L_{2C}} \tag{6.61}$$

This approximation will leave the fifth term in (6.45) as the only significant representative of the Reynolds stress tensor in the governing equation of $\langle V_2\rangle$. Without any further cues available, we retain this term in the simplified form of (6.57).

In summary, the order-of-magnitude analysis of the V_2 equation inside the boundary layer flow over a flat plate has led to the following conclusions. At $Re_1 \gg 1$, the wall-normal acceleration is negligible compared to the stream-wise acceleration. Thus, the $\langle V_2\rangle$ equation reduces merely to a balance of forces acting on a fluid particle in the wall-normal direction. Further, all the viscous forces arising due to the gradients of the wall-normal velocity component are inconsequential in this force balance, leading to a mere balance between the pressure force and the Reynolds stress term, which has a gradient in the wall-normal direction.

$$\frac{\partial\langle p\rangle}{\partial x_2} \approx \frac{\partial\left\langle\rho V_2' V_2'\right\rangle}{\partial x_2} \tag{6.62}$$

We finally assemble the simplified set of governing equations for the mean turbulent flow past a flat plate with $Re_1 \gg 1$. This set consists of Eqs. (6.11), (6.54), and (6.62).

$$\frac{\partial \langle V_1 \rangle}{\partial x_1} + \frac{\partial \langle V_2 \rangle}{\partial x_2} = 0,$$
$$\langle V_1 \rangle \frac{\partial \langle V_1 \rangle}{\partial x_1} + \langle V_2 \rangle \frac{\partial \langle V_1 \rangle}{\partial x_2} = -\frac{1}{\rho}\frac{\partial \langle p \rangle}{\partial x_1} + \nu \frac{\partial^2 \langle V_1 \rangle}{\partial x_2 \partial x_2} - \frac{\partial \left\langle \rho V_1' V_2' \right\rangle}{\partial x_2},$$
$$\frac{\partial p}{\partial x_2} = -\frac{\partial \left\langle \rho V_2' V_2' \right\rangle}{\partial x_2} \tag{6.63}$$

This is a set of three partial differential equations (PDE) with the primary dependent variables being $\langle V_1 \rangle$, $\langle V_2 \rangle$, and $\langle p \rangle$ inside the boundary layer. There are two secondary unknowns appearing in this equation set: $\left\langle -\rho V_1' V_2' \right\rangle$ and $\left\langle -\rho V_2' V_2' \right\rangle$. This equation set is called the *Prandtl boundary layer equation set for a turbulent flow (PBLET)*.

The last equation of (6.63) can be integrated along x_2: between the edge of the boundary layer (denoted by the superscript "edge") and an arbitrary location (with coordinates x_1, x_2, x_3) inside the boundary layer to arrive at

$$\int_{x_2}^{\text{edge}} \frac{\partial \langle p \rangle}{\partial x_2'} dx_2' = -\int_{x_2}^{\text{edge}} \frac{\partial \left\langle \rho V_2' V_2' \right\rangle}{\partial x_2'} dx_2',$$

$$\langle p \rangle = \langle p \rangle^{\text{edge}} + \left\langle \rho V_2' V_2' \right\rangle^{\text{edge}} - \left\langle \rho V_2' V_2' \right\rangle \tag{6.64}$$

Figure (6.2) shows that $\left\langle \rho V_2' V_2' \right\rangle^{\text{edge}} \approx 0$. Thus, (6.70) simplifies to

$$\langle p \rangle = \langle p \rangle^{\text{edge}} - \left\langle \rho V_2' V_2' \right\rangle \tag{6.65}$$

Further, Fig. 6.2 shows that $\left\langle \rho V_2' V_2' \right\rangle$ is restricted to a small fraction of the dynamic pressure of the free stream. Specifically, we observe that the maximum value of $\left\langle V_2' V_2' \right\rangle$ is $0.04 V_o^2$. Thus, at the given stream-wise station in the referred boundary layer [3]

$$\left\langle \rho V_2' V_2' \right\rangle_{\max} \approx 2 \times (0.002) \frac{\rho V_o^2}{2} \tag{6.66}$$

Further, if we have the known condition that the free stream dynamic pressure itself is much smaller than $\langle p \rangle^{\text{edge}}$

$$\frac{\rho V_o^2 / 2}{\langle p \rangle^{\text{edge}}} \ll 1 \tag{6.67}$$

Thus, referring to (6.65), it is plausible to conclude that pressure within the boundary layer at a given stream-wise station (x_1) does not vary significantly with the wall-normal distance (x_2), and its value equals the pressure at the edge of the boundary layer at that stream-wise station.

Since the edge of the boundary layer is actually one of the boundaries of the boundary layer domain described by the PBLET set, the variation of pressure at the edge of the boundary layer can be deemed as a boundary condition of (6.63). Pressure is no longer a variable of the PBLET set (6.69). Accordingly, the pressure gradient term appearing in the $\langle V_1 \rangle$ equation can be expressed as

$$\langle p \rangle = \langle p \rangle^{\text{edge}}, \tag{6.68}$$

where $\langle p \rangle^{\text{edge}}$ is a function of x_1 alone. The PBLET equation set (6.63) can now be expressed more simply as

$$\begin{aligned} \frac{\partial \langle V_1 \rangle}{\partial x_1} + \frac{\partial \langle V_2 \rangle}{\partial x_2} &= 0, \\ \langle V_1 \rangle \frac{\partial \langle V_1 \rangle}{\partial x_1} + \langle V_2 \rangle \frac{\partial \langle V_1 \rangle}{\partial x_2} &= -\frac{1}{\rho}\frac{d \langle p \rangle^{\text{edge}}}{dx_1} + \nu \frac{\partial^2 \langle V_1 \rangle}{\partial x_2 \partial x_2} \\ &\quad - \frac{\partial \left\langle \rho V_1' V_2' \right\rangle}{\partial x_2} \end{aligned} \tag{6.69}$$

The flow domain over a wide flat plate can be perceived to be comprised of two regions which are separated by the curve that describes the edge of the boundary layer $\delta(x_1)$ (Fig. 6.3). The region bounded by the solid plate and the boundary layer edge is indeed the boundary layer region, wherein, by definition, the viscous forces are significant enough to influence the acceleration of a typical fluid particle therein. This demarcation itself implies that the region outside the boundary layer has a negligible influence of viscous forces on the acceleration of the fluid particles. Further, DNS-based evidence (such as Fig. 6.2), suggests that the Reynolds stress components also tend to vanish outside the boundary layer. In this outer region, the flow is not only laminar but also approximately inviscid. Therefore, that region of the flow field must be governed by the steady Euler equation, with the velocity field being 2D and 2C. Further, since the velocity field in the far-upstream region is uniform and lacks any vorticity, the velocity field outside the boundary layer may be approximated to

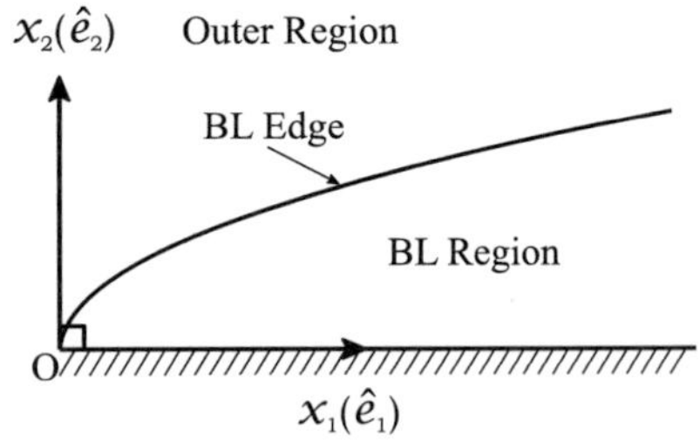

Fig. 6.3 Two regions of the flow field past a flat plate demarcated by the edge of the boundary layer (BL)

be potential (explanation available in [1]). This, in turn, implies that the generalized Bernoulli equation [1] can be applied between any two locations in the region outside the boundary layer. Choosing location A to be on the $x_1\hat{e}_1$ axis in the far-upstream region where both the velocity ($V_o\hat{e}_1$) and pressure (p_o) are known, and choosing B to be located at the edge of the boundary layer at station x_1, we have the following relationship:

$$\langle p\rangle^{\text{edge}}_{(x_1,\delta(x_1))} = p_o + \frac{\rho V_0^2}{2} - \frac{\rho}{2}\langle V\rangle^2_{(x_1,\delta(x_1))} \tag{6.70}$$

where $\langle p\rangle^{\text{edge}}_{(x_1,\delta(x_1))}$ and $\langle V\rangle^2_{(x_1,\delta(x_1))}$ are the pressure and the square of the magnitude of local velocity at the edge of the boundary layer at station x_1 (Fig. 6.3), respectively. The symbol p_o denotes pressure in the far-upstream region of the flow field (Fig. 6.1)

At the edge of the boundary layer, we have,

$$\langle V\rangle^2_{(x_1,\delta(x_1))} = \langle V_1\rangle^2_{(x_1,\delta(x_1))} + \langle V_2\rangle^2_{(x_1,\delta(x_1))} \tag{6.71}$$

However, based on the OM analysis, we are aware that at $Re_1 \gg 1$ (Eqs. 6.42 and 6.50), we have

$$\mathcal{O}\left(\frac{\langle V_2\rangle}{\langle V_1\rangle}\right) = \mathcal{O}\left(\frac{V_{2C}}{V_{1C}}\right) = \mathcal{O}\left(\frac{L_{2C}}{L_{1C}}\right) = \mathcal{O}\left(\frac{1}{\sqrt{Re_1}}\right) \ll 1 \tag{6.72}$$

Thus, we approximate (6.71) to

$$\langle V\rangle^2_{(x_1,\delta(x_1))} \approx \langle V_1\rangle^2_{(x_1,\delta(x_1))} \tag{6.73}$$

Further, by using the definition of $\delta(x_1)$ to express V_1 at the edge of the boundary layer in terms of V_o, Eq. (6.73) is expressed as

$$\langle V\rangle^2_{(x_1,\delta(x_1))} \approx (0.99V_o)^2 \tag{6.74}$$

Thus, (6.70) is expressed as

$$\langle p\rangle^{\text{edge}}_{(x_1,\delta(x_1))} \approx p_o + \frac{0.02\rho(V_o)^2}{2} \tag{6.75}$$

where the RHS is no more a function of x_1. Thus, for a flat plate boundary layer at $Re_1 \gg 1$

$$\frac{dp^{\text{edge}}}{dx_1} \approx 0 \tag{6.76}$$

Accordingly, the PBLET equation set (6.69) is further simplified to

$$\begin{gathered}\frac{\partial\langle V_1\rangle}{\partial x_1} + \frac{\partial\langle V_2\rangle}{\partial x_2} = 0,\\ \langle V_1\rangle\frac{\partial\langle V_1\rangle}{\partial x_1} + \langle V_2\rangle\frac{\partial\langle V_1\rangle}{\partial x_2} = \nu\frac{\partial^2\langle V_1\rangle}{\partial x_2\partial x_2} - \frac{1}{\rho}\frac{\partial\left\langle\rho V_1' V_2'\right\rangle}{\partial x_2}\end{gathered} \tag{6.77}$$

The LHS of the $\langle V_1 \rangle$-equation (6.77) represents the mean stream-wise acceleration of the local fluid particle. The RHS shows the contribution of various external forces (per unit mass) acting on that fluid particle. The first term represents the viscous force per unit mass, and the second term represents the force arising due to the Reynolds stress tensor. Our observation of Fig. 6.2 shows that mean stream-wise velocity monotonically increases as the wall-normal distance increases. Further, we observe that the slope of the mean stream-wise velocity curve decreases with the wall-normal distance. These observations imply that the second derivative of the mean stream-wise velocity in wall-normal distance must be negative. In turn, this means that the net viscous force on the local particle is retarding in nature. This is expected. On the other hand, Fig. 6.2 shows that the wall-normal gradient of the Reynolds stress component $\left\langle -\rho V_1^{'} V_2^{'} \right\rangle$ is positive in the region very close to the wall, and subsequently, at larger wall-normal distances the wall-normal gradient is negative. These observations, in conjugation with the RHS of the $\langle V_1 \rangle$ equation, suggest that in the region close to the wall the force arising due to the Reynolds stress tensor in the mean flow of a flat plate boundary layer causes the local fluid particle to accelerate. However, at larger wall-normal distances, this force changes its nature and causes retardation of a local fluid particle.

6.4 Anatomy of a Flat Plate Turbulent Boundary Layer

The variation of various flow statistics shown in Fig. 6.2 has been plotted against the wall-normal distance, which is normalized by the boundary layer thickness (defined in (6.8)). If measurements are made at different stream-wise stations, the profiles of various expected values included in Fig. 6.2 may show similar trends, but quantitatively they may still differ. However, when these measured expected values as well as the wall-normal distance are appropriately normalized, the consequent profiles of some of these statistics (at different stream-wise stations) tend to collapse and show behavior which seems independent of the stream-wise station of measurement.

The normalization strategy of the wall-normal distance is based on Prandtl's work [4], according to which we define a normalized version (y^+) of the variable x_2 as

$$y^+ = \frac{x_2}{\delta_{\text{viscous}}} \tag{6.78}$$

where δ_{viscous} has dimensions of length and is defined as

$$\delta_{\text{viscous}} = \nu \sqrt{\frac{\rho}{\tau_w}} \tag{6.79}$$

where $\tau_w = \sigma_{12}^{\text{mean}}$ at $x_2 = 0$ at a given x_1. This quantity (τ_w) is called the *wall shear stress* at a chosen x_1 station on the flat plate.

According to (5.20), the mean shear stress ($\sigma_{12}^{\text{mean}}$) in any turbulent flow field is

$$\sigma_{12}^{\text{mean}} = \langle \tau_{12} \rangle + \left\langle -\rho V_1^{'} V_2^{'} \right\rangle \tag{6.80}$$

Here for a Newtonian flow,

$$\langle \tau_{12} \rangle = 2\mu \langle S_{12} \rangle \tag{6.81}$$

Since at $Re_1 \gg 1$, it follows from (6.31) and (6.50)

$$\frac{\partial \langle V_2 \rangle}{\partial x_1} \ll \frac{\partial \langle V_1 \rangle}{\partial x_2} \tag{6.82}$$

Thus, $\langle \tau_{12} \rangle$ (6.81) simplifies to

$$\langle \tau_{12} \rangle \approx \mu \frac{\partial \langle V_1 \rangle}{\partial x_2}. \tag{6.83}$$

At $x_2 = 0$ or, equivalently, at $y^+ = 0$,

$$\begin{aligned} \tau_w = \sigma_{12}^{\text{mean}}|_{x_2=0} &= \langle \tau_{12} \rangle |_{x_2=0} + \left\langle -\rho V_1' V_2' \right\rangle |_{x_2=0} \\ &= \langle \tau_{12} \rangle |_{x_2=0} + 0 && (6.84) \\ &\approx \mu \frac{\partial \langle V_1 \rangle}{\partial x_2}\bigg|_{x_2=0} && (6.85) \end{aligned}$$

Due to the no-slip and the no-penetration boundary conditions at the wall, we have

$$\left\langle -\rho V_1' V_2' \right\rangle \bigg|_{x_2=0} \tag{6.86}$$

Further, the mean stream-wise velocity is normalized to the following form (denoted by u^+)

$$u^+ = \frac{\langle V_1 \rangle}{u_\tau} \tag{6.87}$$

where u_τ is a quantity having the dimensions same as that of velocity and is defined as

$$u_\tau = \sqrt{\frac{\tau_w}{\rho}} \tag{6.88}$$

The quantity u_τ is called the *friction velocity* at the given stream-wise station (x_1).

To better understand the behavior of various flow statistics inside a turbulent layer, we refer to a DNS database described in Schlatter and Örlu [3] (this database itself has been downloaded from the webpage https://www.mech.kth.se/~pschlatt/DATA/ in accordance with the permission provided therein). This downloaded DNS database has been used to generate Figs. 6.4, 6.5, 6.6, 6.7, 6.8, 6.9, and 6.10 of this chapter).

In Fig. 6.4 three representative curves are included corresponding to different stream-wise stations (x_a, x_b, and x_c), such that the Reynolds number based on the local momentum thickness at these three locations is 2000, 3030, and 3970. Both the horizontal and the vertical axes are logarithmic.

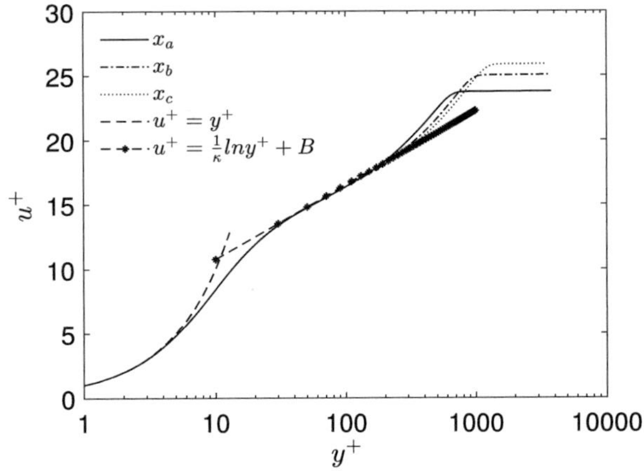

Fig. 6.4 Variation of u^+ versus y^+. Different curves correspond to profiles obtained at different stream-wise stations x_a, x_B, and x_C

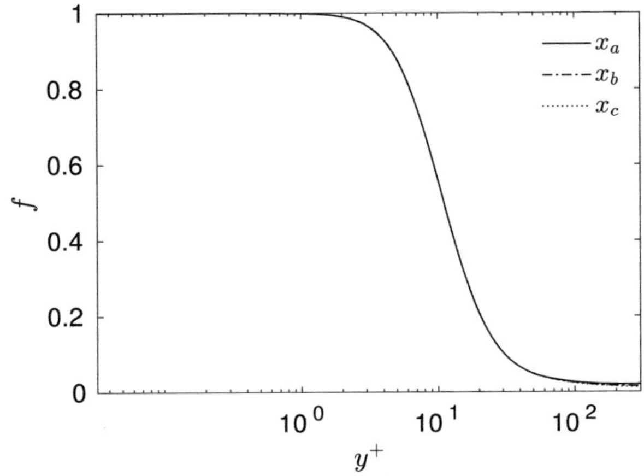

Fig. 6.5 Variation of f (6.91) with y^+ at three different stream-wise stations

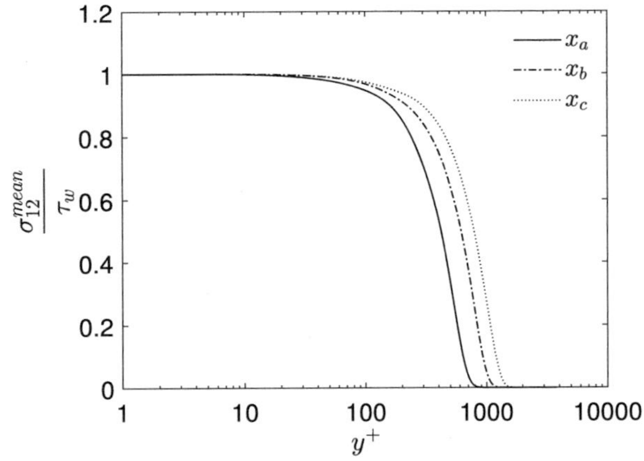

Fig. 6.6 Variation of $\sigma_{12}^{\text{mean}}$ (6.91) with y^+ at three different stream-wise stations. All curves coincide

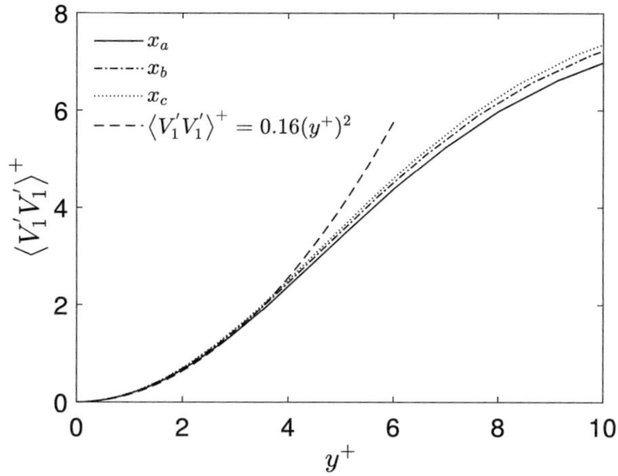

Fig. 6.7 Variation of $\left\langle V_1'V_1' \right\rangle^+$ with y^+ at three different stream-wise stations

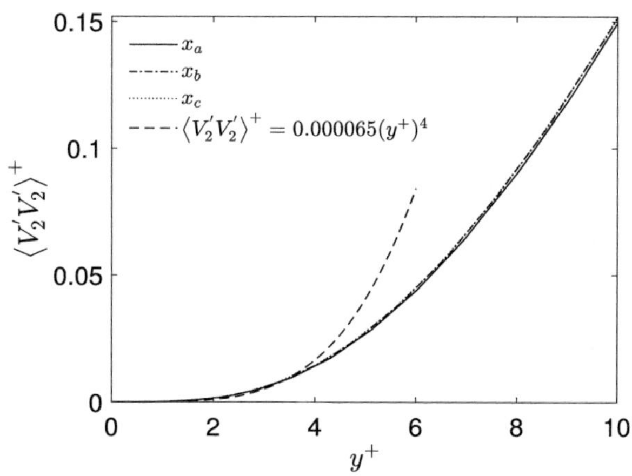

Fig. 6.8 Variation of $\left\langle V_2'V_2' \right\rangle^+$ with y^+ at three different stream-wise stations

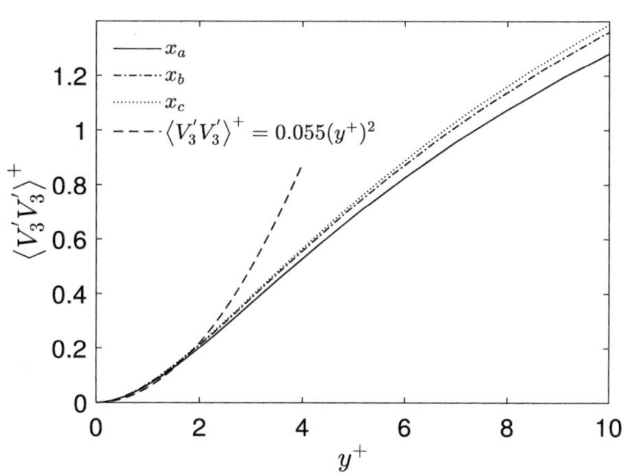

Fig. 6.9 Variation of $\left\langle V_3'V_3' \right\rangle^+$ with y^+ at three different stream-wise stations

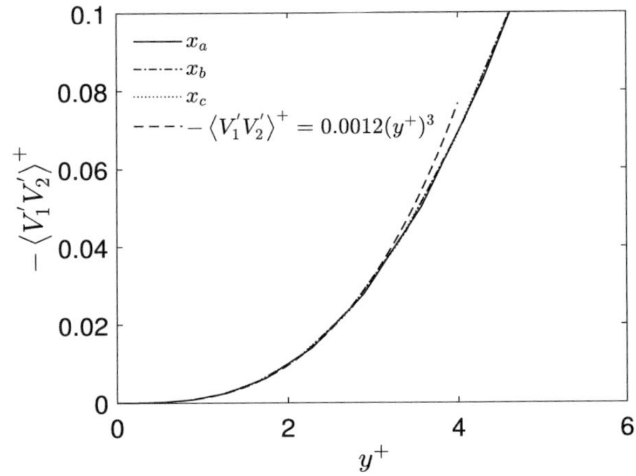

Fig. 6.10 Variation of $-\left\langle V_1'V_2'\right\rangle^+$ with y^+ at three different stream-wise stations

In the range of $y^+ < 5$, at all stream-wise stations, the variation of data seems to follow the curve fit

$$u^+ = y^+ \tag{6.89}$$

This region of the boundary layer is called the *viscous sublayer*. In the region $y^+ > 30$ but $y < 0.2\delta(x_1)$, data from different stream-wise stations collapse on a single curve fit

$$u^+ = \frac{1}{\kappa} lny^+ + B \tag{6.90}$$

where $\kappa = 0.41$ and $B = 5$. This region of the boundary layer wherein the curve fit of (6.90) describes the variation of u^+ with y^+ is called the *log layer*. The relationship (6.90) itself is called the *law of the wall* of the turbulent boundary layer. The region between the log layer and the viscous sublayer ($5 < y^+ < 30$) is called the *buffer layer* and here the data seem to collapse on a curve which merges smoothly with (6.89) on one hand and (6.90) on the other hand. Unlike the viscous sublayer, the buffer layer, or the log layer, beyond $y > 0.2\delta$, the data from different stream-wise stations do not collapse on any common curve. This region of the boundary layer is called the *outer layer*.

Next, we define a quantity f with which we intend to measure the relative importance of the individual contributions from the mean shear viscous stress and the Reynolds shear stress to $\sigma_{12}^{\mathrm{mean}}$.

$$f = \frac{\langle \tau_{12} \rangle}{\langle \tau_{12} \rangle + \left\langle -\rho V_1' V_2' \right\rangle} \tag{6.91}$$

We have defined this quantity such that when f approaches unity, it implies an increased contribution of the viscous stress to $\sigma_{12}^{\mathrm{mean}}$. On the other hand, if f tends to zero, it implies a greater contribution of the Reynolds stress to $\sigma_{12}^{\mathrm{mean}}$. Figure 6.5 shows the variation in f with y^+ inside the boundary layer at three different stream-wise stations. We observe that in the viscous sublayer ($y^+ < 5$), f is almost unity, suggesting that here the Reynolds stress contribution to $\sigma_{12}^{\text{mean}}$ is negligible, and the entire $\sigma_{12}^{\text{mean}}$ is attributable to the mean viscous stress. On the other hand, f almost vanishes in the log layer and beyond, $y^+ > 30$ suggesting that the contribution of the mean viscous stress to $\sigma_{12}^{\text{mean}}$ is negligible, and the entire $\sigma_{12}^{\text{mean}}$ is due to the Reynolds stress itself. Identical trends are observed at all three stream-wise stations. In between, in the buffer layer, f lies in the range of [0, 1] and suggests that there in that layer both the viscous stress and the Reynolds stress contributions to $\sigma_{12}^{\text{mean}}$ are significant.

In Fig. 6.6 we present the variation of $\sigma_{12}^{\text{mean}}/\tau_w$ versus y^+ at three different stream-wise stations. We observe that $\sigma_{12}^{\text{mean}}$ approximately equals the wall stress (τ_w) over an extended region of the boundary layer, which includes the entire viscous sublayer, the entire buffer layer, and a substantial part of the log layer, as well.

Figures 6.4 and 6.6 provide strong evidence about universality in the near-wall region (viscous sublayer, buffer layer, and a substantial part of the log layer) of a turbulent boundary layer at high Re_1. In Chap. 8, we discuss how such universality has been leveraged in turbulence modeling.

6.5 Near-Wall Asymptotic Behavior of the Fluctuating Velocity Vector

Indeed, the no-slip and the no-penetration boundary conditions ensure that all three Cartesian components of the instantaneous velocity vector vanish at the wall ($x_2 = 0$). However, as soon as one steps away from the wall, along the wall-normal direction, the instantaneous velocity vector in a turbulent boundary layer becomes three componential. At this point, our interest is to examine, specifically, what is called the *near-wall asymptotic behavior* of various velocity components. The near-wall asymptotic behavior entails estimating the nature of the dependence of these velocity components on the wall-normal distance (x_2) within the innermost layer (the viscous sublayer) of the boundary layer. For example, with Fig. 6.4 we are already aware that

$$u^+ = y^+ \tag{6.92}$$

which, using (6.78) and (6.87), is recast in terms of $\langle V_1 \rangle$ as

$$\langle V_1 \rangle = \sqrt{\frac{\tau_w}{\rho}}\, x_2 \tag{6.93}$$

which demonstrates that at a fixed stream-wise station, within the viscous sublayer, $\langle V_1 \rangle$ increases linearly with x_2 as one moves away from the wall.

In Fig. 6.7 we present the variation of $\left\langle V_1' V_1' \right\rangle^+$ with y^+ versus y^+, where

$$\left\langle V_1' V_1' \right\rangle^+ = \frac{\left\langle V_1' V_1' \right\rangle}{u_\tau^2}. \tag{6.94}$$

We observe that in the viscous sublayer ($y^+ < 5$), data from different stream-wise stations collapse on the curve fit

$$\left\langle V_1' V_1' \right\rangle^+ = 0.16(y^+)^2 \tag{6.95}$$

which when cast in terms of un-normalized variables result into the following relationship

$$\left\langle V_1' V_1' \right\rangle = 0.16 \left(\frac{\tau_w}{\nu \rho} \right)^2 x_2^2 \tag{6.96}$$

Equation 6.96 shows that at a fixed stream-wise station, within the viscous sublayer, $\left\langle V_1' V_1' \right\rangle$ increases with x_2^2 as one moves away from the wall.

In Fig. 6.8 we present the variation of $\left\langle V_2' V_2' \right\rangle^+$ with y^+, where

$$\left\langle V_2' V_2' \right\rangle^+ = \frac{\left\langle V_2' V_2' \right\rangle}{u_\tau^2} \tag{6.97}$$

We observe that in the viscous sublayer ($y^+ < 5$), data from different stream-wise stations collapse to the curve fit

$$\left\langle V_2' V_2' \right\rangle^+ = 0.000065(y^+)^4 \tag{6.98}$$

which when cast in terms of un-normalized variables result into the following relationship

$$\left\langle V_2' V_2' \right\rangle = 0.000065 \left(\frac{\tau_w^3}{\nu^4 \rho^3} \right) x_2^4 \tag{6.99}$$

Equation 6.99 shows that at a fixed stream-wise station, within the viscous sublayer, $\left\langle V_2' V_2' \right\rangle$ increases with x_2^4 as one moves away from the wall.

In Fig. 6.9 we present the variation of $\left\langle V_3' V_3' \right\rangle^+$ with y^+, where

$$\left\langle V_3' V_3' \right\rangle^+ = \frac{\left\langle V_3' V_3' \right\rangle}{u_\tau^2} \tag{6.100}$$

We observe that in the viscous sublayer ($y^+ < 5$), data from different stream-wise stations collapse to the curve fit

$$\left\langle V_3' V_3' \right\rangle^+ = 0.055(y^+)^2 \tag{6.101}$$

which when cast in terms of un-normalized variables result into the following relationship

$$\left\langle V_3' V_3' \right\rangle = 0.055 \left(\frac{\tau_w}{\nu \rho} \right)^2 x_2^2 \tag{6.102}$$

Equation (6.102) shows that at a fixed stream-wise station, within the viscous sublayer, $\left\langle V_3' V_3' \right\rangle$ increases with x_2^2 as one moves away from the wall.

In Fig. 6.10 we present the variation of $-\left\langle V_1' V_2' \right\rangle^+$ with y^+, where

$$\left\langle V_1' V_2' \right\rangle^+ = \frac{\left\langle V_1' V_2' \right\rangle}{u_\tau^2} \tag{6.103}$$

We observe that in the viscous sublayer ($y^+ < 5$), data from different stream-wise stations collapse to the curve fit

$$-\left\langle V_1' V_2' \right\rangle^+ = 0.0012 (y^+)^3 \tag{6.104}$$

which when cast in terms of un-normalized variables result into the following relationship

$$-\left\langle V_1' V_2' \right\rangle = 0.0012 \left(\frac{\tau_w}{\rho} \right)^{\frac{5}{2}} \frac{x_2^3}{\nu^3} \tag{6.105}$$

Equation (6.105) shows that at a fixed stream-wise station, within the viscous sublayer, $\left\langle V_1' V_2' \right\rangle$ increases with x_2^3 as one moves away from the wall. In Chap. 8, we will highlight how such a knowledge of the near-wall asymptotic behavior of the individual velocity components can possibly be leveraged in modeling the Reynolds stress tensor to achieve closure of the RANS equation set (5.10).

7 Understanding Multiplicity of Length-Scales in Turbulent Flows

In this chapter, we focus on a simple turbulent flow field called *decaying turbulence*. It is a special case of a statistically homogeneous flow field, and it has zero mean velocity at all locations and at all times. Since the mean velocity vector is identical at all locations, the entire mean velocity gradient tensor, too, is zero at all locations.

$$\underline{\nabla}\langle \underline{V}\rangle = \underline{0} \tag{7.1}$$

This implies that the mean continuity equation (5.2) is trivially satisfied at all locations and at all time instants. Further, the mean momentum equation (5.8) simplifies to

$$\frac{\partial \langle V_i\rangle}{\partial t} = -\frac{1}{\rho}\frac{\partial \langle p\rangle}{\partial x_i} \tag{7.2}$$

which merely underlines, the necessary condition that the mean pressure gradient tensor must be zero at all locations and at all times to ensure that $\langle V_i\rangle$ remains zero at all locations and at all times. Thus in decaying turbulence

$$\frac{\partial \langle p\rangle}{\partial x_i} + \frac{\partial p'}{\partial x_i} = \frac{\partial p'}{\partial x_i} \tag{7.3}$$

Further, in this flow field, the RSTE equation (5.43) simplifies to

$$\frac{\partial R_{ij}}{\partial t} = \epsilon_{ij} - \Pi_{ij} \tag{7.4}$$

Since the flow is statistically homogeneous, the spatial gradients of all flow statistics (5.43) vanish. Furthermore, the partial time derivative of R_{ij} may simply be expressed as a total derivative because R_{ij} does not depend on $\underline{X}$.

S. S. Sinha, *Fundamentals of Turbulence and Its Modeling*,
https://doi.org/10.1007/978-3-031-94016-3_7

$$\frac{dR_{ij}}{dt} = \epsilon_{ij} - \Pi_{ij} \tag{7.5}$$

The transport equation of the turbulence kinetic energy

$$\frac{dk}{dt} = -\epsilon \tag{7.6}$$

Equation (7.5) shows that in decaying turbulence, the Reynolds stress tensor evolves under the influence of merely two processes, the dissipation-rate tensor ($\underline{\epsilon}$) and the pressure-strain correlation tensor ($\underline{\Pi}$). The production, molecular diffusion, and the $\underline{\mathcal{T}}$ tensor do not play any role, since all these involve the gradient of the mean velocity vector or the gradient of the expected values of the products of some fluctuating quantities.

Equation (7.6) shows that turbulence kinetic energy evolves under the sole influence of the dissipation-rate tensor. Since ϵ can never be a negative quantity, k decays monotonically in decaying turbulence.

Our primary intent behind examining decaying turbulence is to study the process by which eddies of disparate time and length-scales are generated in a turbulent flow field. The decaying turbulence flow field is apt for such an investigation because it naturally eliminates the influence of the production as well as that of the influence of the inhomogeneous processes of the Reynolds stress tensor (5.41). This makes it easier to develop some deeper insights into the essential *energetics*: how turbulence kinetic energy is converted into heat. Further, this investigation helps us understand the generation of smaller scales of motion in a turbulent flow field.

The initial state of a decaying turbulent flow field is primarily characterized by its initial Reynolds number (Re_λ), which is defined as

$$Re_\lambda = \lambda V_{\text{rms}}/\nu \tag{7.7}$$

where V_{rms} is the root mean squared (rms) of the magnitude of velocity fluctuation in the initial flow field,

$$V_{\text{rms}} = \left\langle \frac{V_i' V_i'}{3} \right\rangle = \frac{2k}{3} \tag{7.8}$$

and k is turbulence kinetic energy per unit mass (5.53). The characteristic length-scale (λ) used in the definition of the Reynolds number (7.7) is called the *Taylor microscale*, and is defined [5] as

$$\lambda = \sqrt{\frac{V_{\text{rms}}^2}{\langle(\partial V_1/\partial x_1)^2\rangle}} \tag{7.9}$$

A detailed examination of a decaying turbulent flow and the influence of the length-scales on the energetics of the flow field requires the availability of the instantaneous velocity field at several time instants and locations. In experiments, one sets up such a decaying turbulent flow field in a wind tunnel wherein the initial turbulence is generated by subjecting a uniform flow to a wire mesh or a grid. The wire mesh generates turbulence kinetic energy in the flow field, which then undergoes a decay process within the central portion considerably distant from the test section walls, such that the influence of the walls of the test section on the decay process is negligible. Such a turbulent flow created using a wire grid inside the test section of a wind tunnel is also called *grid turbulence*. For further details on the experimental methodology, measurement techniques, and computations of flow statistics of grid turbulence, the reader is referred to [6]. In the rest of this chapter, instead, we focus on the details of numerically simulating decaying turbulence (direct numerical simulation). Direct numerical simulation of decaying turbulence is performed by solving the instantaneous Navier-Stokes equation set (3.10) and (3.11) over a cubical domain. Each side of the computational domain is of length 2π. Periodic boundary conditions are imposed on the opposite faces of the domain for the pressure and velocity variables. The instantaneous pressure field is initialized such that it follows the Laplacian equation.

$$\frac{\partial^2 p}{\partial x_i \partial x_i} = -\rho \frac{\partial V_i}{\partial x_j}\frac{\partial V_j}{\partial x_i} \tag{7.10}$$

The instantaneous Navier-Stokes equation set (3.10 and 3.11) is discretized over an appropriately fine computational grid, and time marching is performed with adequately small time steps in order to accurately resolve all scales of motion of the turbulent flow field.

To analyze the initial conditions as well as the evolution of a decaying turbulent flow field at later times, we choose to express the velocity and the pressure field in terms of the *Fourier modes* and the *Fourier amplitudes*. Such expressions also help us better understand the evolution of multiple scales in the flow field. In the next section, we present the mathematical foundation which proves useful in expressing a periodic function using its Fourier description.

7.1 Fourier Description of a Spatially Periodic Function

If ϕ is an Eulerian flow variable which is periodic in each of the three Cartesian directions with its periodicity over length L, then that variable can be expressed in the corresponding complex Fourier description as

$$\phi\left(\underline{X}, t\right) = \sum_{\underline{K}} \widehat{\phi}\left(\underline{K}, t\right) e^{i\underline{K}\cdot\underline{X}} \tag{7.11}$$

where $i = \sqrt{-1}$. The symbol $\underline{K}$ represents an arbitrary *wavenumber vector* defined as

$$\underline{K} = \frac{2\pi}{L}\left(n_1\widehat{e}_1 + n_2\widehat{e}_2 + n_3\widehat{e}_3\right) \tag{7.12}$$

where n_1, n_2 and n_3 are arbitrary integers, and

$$e^{i\underline{K}\cdot\underline{X}} = cos\left(\underline{K}\cdot\underline{X}\right) + isin\left(\underline{K}\cdot\underline{X}\right) \tag{7.13}$$

The complex quantity $e^{i\underline{K}\cdot\underline{X}}$ in (7.11) is called the *Fourier mode* corresponding to the wavenumber vector $\underline{K}$, and $\widehat{\phi}\left(\underline{K}, t\right)$ is called the *Fourier amplitude* corresponding to the wavenumber vector $\underline{K}$. More simply, $\widehat{\phi}(\underline{K}, t)$ is called as the *amplitude of the mode* $e^{i\underline{K}\cdot\underline{X}}$. The summation on the right-hand side of (7.11) is over all possible $\underline{K}$'s that can be generated by all possible integer values of n_1, n_2 and n_3 in accordance with (7.12).

In a flow field which has a periodic spatial distribution of ϕ, potentially, all possible modes (as governed by 7.12) can co-exist. However, whether a particular mode is present or not in the flow field depends entirely on the corresponding Fourier amplitude. If for a chosen $\underline{K}$, $\widehat{\phi}(K, t) = 0$, it means that the mode $e^{i\underline{K}\cdot\underline{X}}$ is absent in the flow field. Since the possible wavenumber vectors are generated by discrete values of n_1, n_2, and n_3, the set of $\underline{K}$ vectors is also a discrete set. On the other hand, $\widehat{\phi}\left(\underline{K}, t\right)$ is a continuous function of time.

The existence of a mode $e^{i\underline{K}\cdot\underline{X}}$ in the field $\phi(\underline{X})$ can be associated with the existence of a length-scale (denoted by the symbol l_K) such that

$$l_K = \frac{2\pi}{K} \tag{7.14}$$

where the symbol K denotes the magnitude of the wavenumber vector $\underline{K}$.

$$K = |\underline{K}| \tag{7.15}$$

The association of this length-scale l_K to K is illustrated in Fig. 7.1.

In this figure, a particular wavenumber vector has been chosen for the purpose of illustration:

$$\underline{K} = \frac{2\pi}{L}(3\widehat{e_1} + 2\widehat{e_2}) \tag{7.16}$$

Correspondingly shaded contours of the quantity $cos(\underline{K}\cdot\underline{X})$ have been plotted with the two axes representing the Cartesian axes $x_1\left(\widehat{e_1}\right)$ and $x_2\left(\widehat{e_2}\right)$ in physical space. We observe that the variation of $cos(\underline{K}\cdot\underline{X})$ represents a wave in the direction perpendicular to the direction of the chosen wavenumber vector (7.16). It can be verified that the contour pattern repeats itself over a length l_K, which can be computed using (7.14). This length (l_K) is alternatively referred to as the *wavelength* associated with the wavenumber vector $\underline{K}$. Based on the general understanding that emerges from the illustration of Fig. 7.1, we infer: the existence of a mode $e^{i\underline{K}\cdot\underline{X}}$ in the field $\phi(\underline{X})$ implies the existence of a corresponding length-scale l_K (7.14) in the field of $\phi(\underline{X})$.

Since the length-scale itself is a scalar quantity, it is quite possible that two wavenumber vectors have different orientations and still have the same magnitude (K). Both such wavenumber vectors are indeed associated with the same length-scale l_K (7.14).

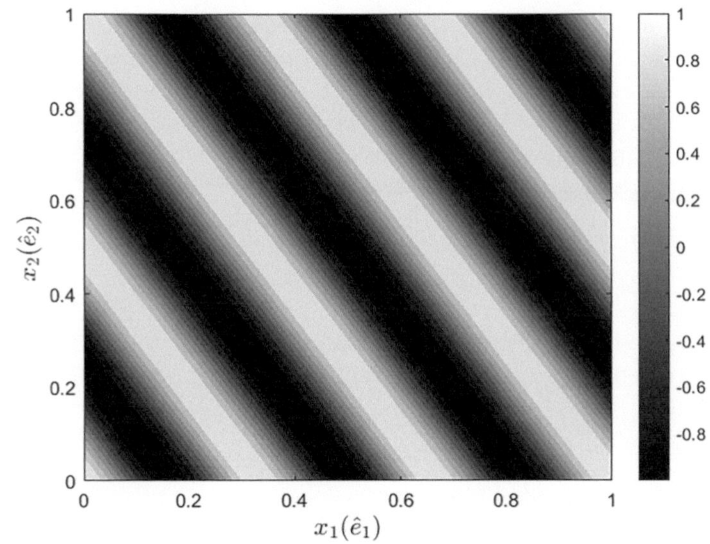

Fig. 7.1 Variation of $cos(\underline{K} \cdot \underline{X})$ in the physical space with $K = \frac{2\pi}{L}(3\widehat{e_1} + 2\widehat{e_2})$, and $L = 1$

In Chap. 2, we discussed how the presence of multiple eddies with different time and length-scales introduces those time and length-scales in the expression of the velocity of a fluid particle in the flow field. Since in a periodic flow field (2.35), the presence of various Fourier modes is also related to the introduction of new length-scales (Fig. 7.1), in turbulence literature, a phenomenological correspondence (without necessarily referring to a mathematically rigorous connection) is often made between the length-scale associated with a Fourier mode (7.14) to the presence of eddies of the diameter of that length-scale in the velocity field. Thus, if a mode with wavenumber vector $\underline{K}$ exists in the flow field, we say that eddies of characteristic length-scale l_{eddy} exist in the flow field such that

$$\mathcal{O}(l_{eddy}) = \mathcal{O}\left(\frac{2\pi}{K}\right) \tag{7.17}$$

where K denotes the magnitude of the wavenumber vector.

We can prove that, in general,

$$\widehat{\phi}\left(\underline{K}\right) = \frac{1}{L^3}\int_0^L\int_0^L\int_0^L \phi\left(\underline{x}\right)e^{-i\underline{K}\cdot\underline{X}}dx_1dx_2dx_3 = \left\langle \phi\left(\underline{x}\right)e^{-i\underline{K}\cdot\underline{X}}\right\rangle_{\mathcal{V}} \tag{7.18}$$

where the argument t has been omitted merely for algebraic brevity. The symbols x_1, x_2, and x_3 denote the Cartesian coordinates of the position vector $\underline{X}$ of an arbitrary location within the cubical domain with each edge length being L.

$$\underline{X} = x_i\widehat{e_i} \tag{7.19}$$

The symbols dx_1, dx_2, and dx_3 are the dimensions of a Cartesian infinitesimal control volume with its centroid at (x_1, x_2, x_3). Equation (7.18) indeed represents the weighted volume average of the flow variable ϕ over the entire cubical domain. The weighting factor in this integration is the complex conjugate of the mode function $e^{i\underline{K}\cdot\underline{X}}$. For algebraic brevity, we use the following symbol to denote the volume-averaging procedure.

$$\langle (\ldots) \rangle_{\mathcal{V}} = \frac{1}{L^3} \int_0^L \int_0^L \int_0^L (\ldots) \, dx_1 dx_2 dx_3 \tag{7.20}$$

We will demonstrate the proof of (7.18). However, before we can do that, we need some useful algebraic properties of the volume-averaging operator (7.20). It is easy to show that

$$\frac{1}{L} \int_0^L e^{\frac{i2\pi n_1 x_1}{L}} dx_1 = 1 \text{ if } n_1 = 0 \tag{7.21}$$

On the other hand, if $n_1 \neq 0$,

$$\begin{aligned}
\frac{1}{L} \int_0^L e^{\frac{i2\pi n_1 x_1}{L}} dx_1 &= \frac{1}{L} \int_0^L \left[cos\left(\frac{2\pi n_1 x_1}{L}\right) + isin\left(\frac{2\pi n_1 x_1}{L}\right)\right] dx_1 \\
&= \frac{1}{2\pi n_1 L} \left[sin\left(\frac{2\pi n_1 x_1}{L}\right) - isin\left(\frac{2\pi n_1 x_1}{L}\right)\right]_0^L \\
&= \frac{1}{2\pi n_1} \{(\sin(2\pi n_1) - \sin(0)) - i(\cos(2\pi n_1) - \cos(0))\} \\
&= \frac{1}{2\pi n_1} \{(0 - 0) - i(1 - 1)\} \\
&= 0
\end{aligned} \tag{7.22}$$

Using (7.21) and (7.22) we now simplify the following expression

$$\psi = \frac{1}{L^3} \int_0^L \int_0^L \int_0^L e^{i\underline{K}\cdot\underline{X}} e^{-i\underline{K}'\cdot\underline{X}} dx_1 dx_2 dx_3$$

where

$$\begin{aligned}
\underline{K} &= K_i \widehat{e_i} = (n_1 \widehat{e_1} + n_2 \widehat{e_2} + n_3 \widehat{e_3}) \left(\frac{2\pi}{L}\right) \\
\underline{K}' &= K_i' \widehat{e_i} = \left(n_1' \widehat{e_1} + n_2' \widehat{e_2} + n_3' \widehat{e_3}\right) \left(\frac{2\pi}{L}\right)
\end{aligned} \tag{7.23}$$

where n_1, n_2, n_3, and n_1', n_2', n_3' are arbitrary integers.

$$
\begin{aligned}
\psi &= \frac{1}{L^3}\int_0^L\int_0^L\int_0^L e^{i\underline{K}\cdot\underline{X}}e^{-i\underline{K}'\cdot\underline{X}}dx_1dx_2dx_3 \\
&= \frac{1}{L^3}\int_0^L\int_0^L\int_0^L e^{i(\underline{K}-\underline{K}')\cdot\underline{X}}dx_1dx_2dx_3 \\
&= \frac{1}{L^3}\int_0^L\int_0^L\int_0^L e^{i(K_1-K_1')x_1}e^{i(K_2-K_2')x_2}e^{i(K_3-K_3')x_3}dx_1dx_2dx_3 \\
&= \left[\frac{1}{L}\int_0^L e^{i(K_1-K_1')x_1}dx_1\right]\left[\frac{1}{L}\int_0^L e^{i(K_2-K_2')x_2}dx_2\right]\left[\frac{1}{L}\int_0^L e^{i(K_3-K_3')x_3}dx_3\right] \\
&= \left[\frac{1}{L}\int_0^L e^{i\frac{2\pi}{L}(n_1-n_1')x_1}dx_1\right]\left[\frac{1}{L}\int_0^L e^{i\frac{2\pi}{L}(n_2-n_2')x_2}dx_2\right]\times \\
&\quad \left[\frac{1}{L}\int_0^L e^{i\frac{2\pi}{L}(n_3-n_3')x_3}dx_3\right]
\end{aligned}
\tag{7.24}
$$

Clearly, if $\underline{K}' = \underline{K}$ (which means $n_1 = n_1'$, $n_2 = n_2'$, $n_3 = n_3'$), we use (7.21) in (7.24) to conclude

$$\frac{1}{L^3}\int_0^L\int_0^L\int_0^L e^{i\underline{K}\cdot\underline{X}}e^{-i\underline{K}'\cdot\underline{X}}dx_1dx_2dx_3 = 1 \tag{7.25}$$

On the other hand, if $\underline{K}' \neq \underline{K}$ (which means at least one of the following three inequalities $n_1 \neq n_1'$, $n_2 \neq n_2'$, $n_3 \neq n_3'$ holds good), we use (7.22) in (7.24) to conclude

$$\frac{1}{L^3}\int_0^L\int_0^L\int_0^L e^{i\underline{K}\cdot\underline{X}}e^{-i\underline{K}'\cdot\underline{X}}dx_1dx_2dx_3 = 0 \tag{7.26}$$

The identities (7.25) and (7.26) are expressed in a more compact manner as

$$\left\langle e^{i\underline{K}\cdot\underline{X}}e^{-i\underline{K}'\cdot\underline{X}}\right\rangle_{\mathcal{V}} = \begin{cases} 1 & \text{if } \underline{K} = \underline{K}' \\ 0 & \text{if } \underline{K} \neq \underline{K}' \end{cases} \tag{7.27}$$

where the symbol defined in (7.20) for the volume-averaging operation has been employed.

Further, we define a new symbol $\delta_{\underline{K},\underline{K}'}$ to express (7.27) as

$$\left\langle e^{i\underline{K}\cdot\underline{X}}e^{-i\underline{K}'\cdot\underline{X}}\right\rangle_{\mathcal{V}} = \delta_{\underline{K},\underline{K}'} \tag{7.28}$$

where

$$\delta_{\underline{K},\underline{K}'} = \begin{cases} 1 & \text{if } \underline{K} = \underline{K}' \\ 0 & \text{if } \underline{K} \neq \underline{K}' \end{cases} \tag{7.29}$$

We are now in a position to present the proof of (7.18). We start with the expression of (7.11).

$$\phi\left(\underline{X}, t\right) = \sum_{\underline{K}} \widehat{\phi}\left(\underline{K}, t\right) e^{i\underline{K}\cdot\underline{X}} \tag{7.30}$$

We multiply both sides by $e^{-i\underline{K}'\cdot\underline{X}}$ of (7.30) leads to

$$\phi\left(\underline{X}, t\right) e^{-i\underline{K}'\cdot\underline{X}} = \sum_{\underline{K}} \widehat{\phi}\left(\underline{K}, t\right) e^{i\underline{K}\cdot\underline{X}} e^{-i\underline{K}'\cdot\underline{X}} \tag{7.31}$$

We now subject both sides of (7.31) to volume averaging

$$\begin{aligned}\left\langle \phi\left(\underline{X}, t\right) e^{-i\underline{K}'\cdot\underline{X}} \right\rangle_{\mathcal{V}} &= \left\langle \sum_{\underline{K}} \widehat{\phi}\left(\underline{K}, t\right) e^{i\underline{K}\cdot\underline{X}} e^{-i\underline{K}'\cdot\underline{X}} \right\rangle_{\mathcal{V}} \\ &= \sum_{\underline{K}} \widehat{\phi}\left(\underline{K}, t\right) \left\langle e^{i\underline{K}\cdot\underline{X}} e^{-i\underline{K}'\cdot\underline{X}} \right\rangle_{\mathcal{V}}\end{aligned} \tag{7.32}$$

Using (7.29) in (7.32) leads to

$$\left\langle \phi\left(\underline{X}, t\right) e^{-i\underline{K}'\cdot\underline{X}} \right\rangle_{\mathcal{V}} = \sum_{\underline{K}} \widehat{\phi}\left(\underline{K}, t\right) \left\langle e^{i\underline{K}\cdot\underline{X}} e^{-i\underline{K}'\cdot\underline{X}} \right\rangle_{\mathcal{V}} = \sum_{\underline{K}} \widehat{\phi}\left(\underline{K}, t\right) \delta_{\underline{K},\underline{K}'} \tag{7.33}$$

The RHS of (7.33) is a summation over all possible modes. However, in that summation, all terms, except the one for which $\underline{K} = \underline{K}'$, vanish. Thus, (7.33) simplifies to

$$\left\langle \phi\left(\underline{X}, t\right) e^{-i\underline{K}'\cdot\underline{X}} \right\rangle_{\mathcal{V}} = \widehat{\phi}\left(\underline{K}', t\right)$$

which is identical to the relationship we listed in (7.18).

Since in this chapter, we will be employing the Fourier expression of various flow variables quite extensively, for algebraic brevity, we introduce a new symbolic operator $\mathcal{F}_{\underline{K}}$, which we define as

$$\mathcal{F}_{\underline{K}}\left\{\phi\left(\underline{X}\right)\right\} = \widehat{\phi}\left(\underline{K}\right) \tag{7.34}$$

Further, we list some of the useful properties of the $\mathcal{F}_{\underline{K}}$ operator, which we will employ in this chapter.

1. $\mathcal{F}_{\underline{K}}\left\{\phi_1(\underline{X}) + \phi_2(\underline{X})\right\} = \widehat{\phi_1}(\underline{K}) + \widehat{\phi_2}(\underline{K})$ where $\phi_1(\underline{X})$ and $\phi_2(\underline{X})$ are two real-valued functions.

Proof:

$$\begin{aligned}\mathcal{F}_{\underline{K}}\left\{\phi_1(\underline{X}) + \phi_2(\underline{X})\right\} &= \left\langle (\phi_1(\underline{X}) + \phi_2(\underline{X})) e^{-i\underline{K}\cdot\underline{X}} \right\rangle_{\mathcal{V}} \\ &= \left\langle \{\phi_1(\underline{X}) e^{-i\underline{K}\cdot\underline{X}} + \phi_2(\underline{X}) e^{-i\underline{K}\cdot\underline{X}}\} \right\rangle_{\mathcal{V}}\end{aligned} \tag{7.35}$$

Since the volume-averaging procedure (7.20) is essentially an integration process, it naturally distributes over the sum of the two functions in (7.35), leading to

$$\begin{aligned}\mathcal{F}_{\underline{K}}\left\{\phi_1(\underline{X})+\phi_2(\underline{X})\right\} &= \left\langle\{\phi_1(\underline{X})+\phi_2(\underline{X})\}e^{-i\underline{K}\cdot\underline{X}}\right\rangle_{\mathcal{V}}\\ &= \left\langle\phi_1(\underline{X})e^{-i\underline{K}\cdot\underline{X}}\right\rangle_{\mathcal{V}}+\left\langle\phi_2(\underline{X})e^{-i\underline{K}\cdot\underline{X}}\right\rangle_{\mathcal{V}}\end{aligned} \tag{7.36}$$

Using (7.18) in (7.36) leads to

$$\mathcal{F}_{\underline{K}}\left\{\phi_1(\underline{X})+\phi_2(\underline{X})\right\}=\widehat{\phi}_1(\underline{K})+\widehat{\phi}_2(\underline{K}). \tag{7.37}$$

2. $\mathcal{F}_{\underline{K}}\left\{\frac{\partial\phi(\underline{X})}{\partial x_i}\right\}=iK_i\widehat{\phi}\left(\underline{K}\right)$

Proof:

$$\begin{aligned}\mathcal{F}_{\underline{K}}\left\{\frac{\partial\phi(\underline{X})}{\partial x_i}\right\} &= \left\langle\frac{\partial\phi}{\partial x_i}e^{-i\underline{K}\cdot\underline{X}}\right\rangle_{\mathcal{V}}=\left\langle\frac{\partial}{\partial x_i}\left[\sum_{\underline{K}'}\widehat{\phi}\left(\underline{K}'\right)e^{i\underline{K}'\cdot\underline{X}}\right]e^{-i\underline{K}\cdot\underline{X}}\right\rangle_{\mathcal{V}}\\ &= \left\langle\sum_{\underline{K}'}\left[\widehat{\phi}\left(\underline{K}'\right)\frac{\partial}{\partial x_i}\left(e^{i\underline{K}'\cdot\underline{X}}\right)\right]e^{-i\underline{K}\cdot\underline{X}}\right\rangle_{\mathcal{V}}\\ &= \left\langle\sum_{\underline{K}'}\left[\widehat{\phi}\left(\underline{K}'\right)\frac{\partial}{\partial x_i}\left(e^{iK_m'x_m}\right)\right]e^{-i\underline{K}\cdot\underline{X}}\right\rangle_{\mathcal{V}}\\ &= \left\langle\sum_{\underline{K}'}\left[\widehat{\phi}\left(\underline{K}'\right)\left(e^{iK_m'x_m}iK_p{}'\frac{\partial x_p}{\partial x_i}\right)\right]e^{-i\underline{K}\cdot\underline{X}}\right\rangle_{\mathcal{V}}\\ &= \left\langle\sum_{\underline{K}'}\left[\widehat{\phi}\left(\underline{K}'\right)\left(e^{iK_m'x_m}iK_p{}'\delta_{pi}\right)\right]e^{-i\underline{K}\cdot\underline{X}}\right\rangle_{\mathcal{V}}\\ &= \left\langle\sum_{\underline{K}'}\left[\widehat{\phi}\left(\underline{K}'\right)\left(e^{i\underline{K}'\cdot\underline{X}}iK_i{}'\right)\right]e^{-i\underline{K}\cdot\underline{X}}\right\rangle_{\mathcal{V}}\\ &= \sum_{\underline{K}'}\left[\widehat{\phi}\left(\underline{K}'\right)(iK_i{}')\left\langle e^{i\underline{K}'\cdot\underline{X}}e^{-i\underline{K}\cdot\underline{X}}\right\rangle_{\mathcal{V}}\right]\\ &= \sum_{\underline{K}'}\left[\widehat{\phi}\left(\underline{K}'\right)(iK_i{}')\delta_{\underline{K}\,\underline{K}'}\right]=iK_i\widehat{\phi}\left(\underline{K}\right)\end{aligned} \tag{7.38}$$

where $\frac{\partial x_p}{\partial x_i}=\delta_{pi}$ has been employed.

Using (7.11) in (7.38) leads to the following conclusion,

$$\mathcal{F}_{\underline{K}}\left\{\frac{\partial \phi(\underline{X})}{\partial x_i}\right\} = i K_i \widehat{\phi}\left(\underline{K}\right). \tag{7.39}$$

3. $\mathcal{F}_{\underline{K}}\left\{\frac{\partial \phi(\underline{X})}{\partial t}\right\} = \frac{d\widehat{\phi}(\underline{K})}{dt}$

Proof:

$$\begin{aligned}
\mathcal{F}_{\underline{K}}\left\{\frac{\partial \phi(\underline{X})}{\partial t}\right\} &= \left\langle \frac{\partial \phi}{\partial t} e^{-i\underline{K}\cdot\underline{X}}\right\rangle_{\mathcal{V}} = \left\langle \frac{\partial}{\partial t}\left[\sum_{\underline{K}'} \widehat{\phi}\left(\underline{K}', t\right) e^{i\underline{K}'\cdot\underline{X}}\right] e^{-i\underline{K}\cdot\underline{X}}\right\rangle_{\mathcal{V}} \\
&= \left\langle \sum_{\underline{K}'}\left[\frac{\partial \widehat{\phi}\left(\underline{K}'\right)}{\partial t}\left(e^{i\underline{K}'\cdot\underline{X}}\right)\right] e^{-i\underline{K}\cdot\underline{X}}\right\rangle_{\mathcal{V}} \\
&= \sum_{\underline{K}'}\left[\frac{\partial \widehat{\phi}\left(\underline{K}'\right)}{\partial t}\left\langle e^{i\underline{K}'\cdot\underline{X}} e^{-i\underline{K}\cdot\underline{X}}\right\rangle_{\mathcal{V}}\right] \\
&= \sum_{\underline{K}'}\left[\frac{\partial \widehat{\phi}\left(\underline{K}'\right)}{\partial t}\delta_{\underline{K}',\underline{K}}\right] \\
&= \frac{\partial \widehat{\phi}\left(\underline{K}\right)}{\partial t}
\end{aligned}$$

The wavenumber vector ($\underline{K}$) appearing as in the argument of $\widehat{V}_i(\underline{K})$ is merely a discrete tag and, thus $\widehat{V}_i(\underline{K})$ is actually a continuous function of time alone. Thus we can express the partial derivative with time as a total derivative.

$$\mathcal{F}_{\underline{K}}\left\{\frac{\partial \phi(\underline{X})}{\partial t}\right\} = \frac{d\hat{\phi}(\underline{K}, t)}{dt} \tag{7.40}$$

4. $\widehat{\phi}(\underline{K}) = \widehat{\phi}^*(-\underline{K})$ where $\widehat{\phi}^*(\underline{K})$ is the complex conjugate of $\widehat{\phi}(\underline{K})$ and $\phi(\underline{X})$ is a real-valued function.

Proof:

$$\widehat{\phi}(\underline{K}) = \left\langle \phi(\underline{X}) e^{-i\underline{K}\cdot\underline{X}}\right\rangle_{\mathcal{V}} \tag{7.41}$$

We take the complex conjugate of both sides of (7.41)

$$\begin{aligned}\widehat{\phi}^{*}(\underline{K}) &= \left\langle \phi(\underline{X}) e^{-i\underline{K}\cdot\underline{X}} \right\rangle_{\mathcal{V}}^{*} \\ &= \left\langle [\phi(\underline{X})]^{*} \left[e^{-i\underline{K}\cdot\underline{X}} \right]^{*} \right\rangle_{\mathcal{V}} \\ &= \left\langle [\phi(\underline{X})] \left[e^{i(\underline{K})\cdot\underline{X}} \right] \right\rangle_{\mathcal{V}}\end{aligned} \tag{7.42}$$

Setting $\underline{K}^{'} = -\underline{K}$, (7.42) is expressed as

$$\widehat{\phi}^{*}(-\underline{K}^{'}) = \left\langle \phi(\underline{X}) e^{-i\underline{K}^{'}\cdot\underline{X}} \right\rangle_{\mathcal{V}} = \widehat{\phi}(\underline{K}^{'}) \tag{7.43}$$

which can be equivalently expressed as

$$\widehat{\phi}^{*}(-\underline{K}^{'}) = \left\langle \phi(\underline{X}) e^{-i\underline{K}^{'}\cdot\underline{X}} \right\rangle_{\mathcal{V}} = \widehat{\phi}(\underline{K}^{'}) \tag{7.44}$$

without any loss of generality at this stage, we substitute $\underline{K}'$ by $\underline{K}$.

$$\widehat{\phi}^{*}(-\underline{K}) = \widehat{\phi}(\underline{K}) \tag{7.45}$$

5.

$$\langle \phi_1(\underline{X})\phi_2(\underline{X}) \rangle_{\mathcal{V}} = \sum_{\underline{K}} \widehat{\phi}_1(\underline{K}) \widehat{\phi}_2^{*}(\underline{K}) \tag{7.46}$$

where $\phi_1(\underline{X})$ and $\phi_2(\underline{X})$ are two real-valued functions.

Proof: We start with the product of the two functions $\phi_1(\underline{X})$ and $\phi_2(\underline{X})$

$$\begin{aligned}\phi_1(\underline{X})\phi_2(\underline{X}) &= \left[\sum_{\underline{K}^{'}} \widehat{\phi}_1(\underline{K}^{'}) e^{i\underline{K}^{'}\cdot\underline{X}} \right] \left[\sum_{\underline{K}^{''}} \widehat{\phi}_2(\underline{K}^{''}) e^{i\underline{K}^{''}\cdot\underline{X}} \right] \\ &= \sum_{\underline{K}^{'}} \sum_{\underline{K}^{''}} \left[\widehat{\phi}_1(\underline{K}^{'}) \widehat{\phi}_2(\underline{K}^{''}) e^{i\underline{K}^{'}\cdot\underline{X}} e^{i\underline{K}^{''}\cdot\underline{X}} \right]\end{aligned} \tag{7.47}$$

subjecting both sides of (7.47) to volume averaging leads to

$$\begin{aligned}\langle \phi_1(\underline{X})\phi_2(\underline{X}) \rangle_{\mathcal{V}} &= \left\langle \sum_{\underline{K}^{'}} \sum_{\underline{K}^{''}} \left[\widehat{\phi}_1(\underline{K}^{'}) \widehat{\phi}_2(\underline{K}^{''}) e^{i\underline{K}^{'}\cdot\underline{X}} e^{i\underline{K}^{''}\cdot\underline{X}} \right] \right\rangle_{\mathcal{V}} \\ &= \sum_{\underline{K}^{'}} \sum_{\underline{K}^{''}} \left[\widehat{\phi}_1(\underline{K}^{'}) \widehat{\phi}_2(\underline{K}^{''}) \left\langle e^{i\underline{K}^{'}\cdot\underline{X}} e^{i\underline{K}^{''}\cdot\underline{X}} \right\rangle_{\mathcal{V}} \right]\end{aligned} \tag{7.48}$$

Using (7.28) in (7.48) leads to

$$\langle \phi_1(\underline{X})\phi_2(\underline{X})\rangle_{\mathcal{V}} = \sum_{\underline{K}'}\sum_{\underline{K}''}\left[\widehat{\phi_1}(\underline{K}')\widehat{\phi_2}(\underline{K}'')\delta_{\underline{K}',-\underline{K}''}\right] \tag{7.49}$$

The appearance of the Kronecker delta symbol in (7.49) implies that the expression can be simplified by (a) discarding the summation operation over $\underline{K}''$ from the RHS of (7.49) and then (b) substituting $\underline{K}''$ by $-\underline{K}'$ in the remaining part of the expression

$$\langle \phi_1(\underline{X})\phi_2(\underline{X})\rangle_{\mathcal{V}} = \sum_{\underline{K}'}\left[\widehat{\phi_1}(\underline{K}')\widehat{\phi_2}(-\underline{K}')\right] \tag{7.50}$$

Further using in (7.44) in (7.50) leads to

$$\langle \phi_1(\underline{X})\phi_2(\underline{X})\rangle_{\mathcal{V}} = \sum_{\underline{K}'}\left[\widehat{\phi_1}(\underline{K}')\widehat{\phi_2^*}(\underline{K}')\right] \tag{7.51}$$

which can equivalently be expressed as

$$\langle \phi_1(\underline{X})\phi_2(\underline{X})\rangle_{\mathcal{V}} = \sum_{\underline{K}}\left[\widehat{\phi_1}(\underline{K})\widehat{\phi_2^*}(\underline{K})\right] \tag{7.52}$$

Equation (7.52) is called the *Parseval's theorem.*

7.2 Spectral Density Functions of Turbulence Kinetic Energy and Its Dissipation Rate

In a direct numerical simulation of decaying turbulence, the imposed boundary conditions on the opposite faces of the cubical domain are periodic. Further, such a simulation is initiated with a periodic velocity field with the periodicity being $L = 2\pi$ in each of the three Cartesian directions. Under these conditions, the velocity and the pressure fields continue to be periodic at all later times. However, with time, the composition of the velocity and the pressure field changes in terms of the existence of various Fourier modes of these variables.

A decaying turbulent flow field is an example of a statistically homogeneous flow field. Thus, volume averaging of flow variables or their functions can be leveraged to estimate their expected values (4.62, Chap. 5). A direct numerical simulation of decaying turbulence provides extensive data on various flow variables at all grid points in the cubical domain and at every discrete (but small) time step of the simulation. This database can be employed to estimate the mean velocity and pressure fields at any chosen instant, as

$$\langle \underline{V}\rangle = \langle \underline{V}\rangle_{\mathcal{V}} \text{ and } \langle p\rangle = \langle p\rangle_{\mathcal{V}} \tag{7.53}$$

where $\mathcal{V}$ is the cubical domain with each edge being of length 2π and $\langle\phi\rangle_{\mathcal{V}}$ is the volume averaged value of ϕ over the cubical domain (see Eq. 4.62). Similarly, the turbulence kinetic at any time instant in the flow field is estimated as

$$k = \frac{1}{2}\left\langle (V_i - \langle V_i\rangle_{\mathcal{V}})(V_i - \langle V_i\rangle_{\mathcal{V}})\right\rangle_{\mathcal{V}}$$

DNS databases of decaying turbulence do show that

$$\frac{|\langle \underline{V}\rangle|}{k^{1/2}} \approx 0 \tag{7.54}$$

where $|\langle \underline{V}\rangle|$ is the magnitude of the mean velocity vector. Thus, in a decaying turbulent flow field

$$V_i = \langle V\rangle + V_i' \approx V_i' \tag{7.55}$$

Thus,

$$\widehat{V}_i(\underline{K}) = \mathcal{F}_{\underline{K}}\left\{V_i^{'}\right\} = \mathcal{F}_{\underline{K}}\left\{V_i\right\}. \tag{7.56}$$

Since the individual velocity components are periodic at all time instants in all three Cartesian directions, using (7.52) and (7.56), we express the instantaneous turbulence kinetic energy (k) as a summation of appropriate Fourier amplitudes over the space of wavenumber vectors:

$$\begin{aligned} k &= \frac{1}{2}\left\langle V_i^{'} V_i^{'}\right\rangle_{\mathcal{V}} = \frac{1}{2}\left\langle V_1^{'}V_1^{'} + V_2^{'}V_2^{'} + V_3^{'}V_3^{'}\right\rangle_{\mathcal{V}} \\ &= \frac{1}{2}\sum_{\underline{K}}[\widehat{V}_1(\underline{K})\widehat{V}_1^*(\underline{K})] + \frac{1}{2}\sum_{\underline{K}}[\widehat{V}_2(\underline{K})\widehat{V}_2^*(\underline{K})] + \frac{1}{2}\sum_{\underline{K}}[\widehat{V}_3(\underline{K})\widehat{V}_3^*(\underline{K})] \\ &= \frac{1}{2}\sum_{\underline{K}}[\widehat{V}_i(\underline{K})\widehat{V}_i^*(\underline{K})] \end{aligned} \tag{7.57}$$

Based on the relationship (7.57), we define the *spectral density function* of k, $(E(K))$ as

$$E(K) = \lim_{\Delta K\to 0}\frac{1}{\Delta K}\frac{1}{2}\sum_{\underline{K}^{'}}\widehat{V}_i(\underline{K}^{'})\widehat{V}_i^*(\underline{K}^{'}) \tag{7.58}$$

where the summation on the RHS of (7.58) is over all those wavenumber vectors $\underline{K}^{'}$ such that

$$K \le |\underline{K}^{'}| < K + \Delta K \tag{7.59}$$

Here K is a scalar quantity (≥ 0) and ΔK is a small independent increment in the value of K. It follows from the definition of E(K) that

$$\int_0^\infty E(K)dK = k. \tag{7.60}$$

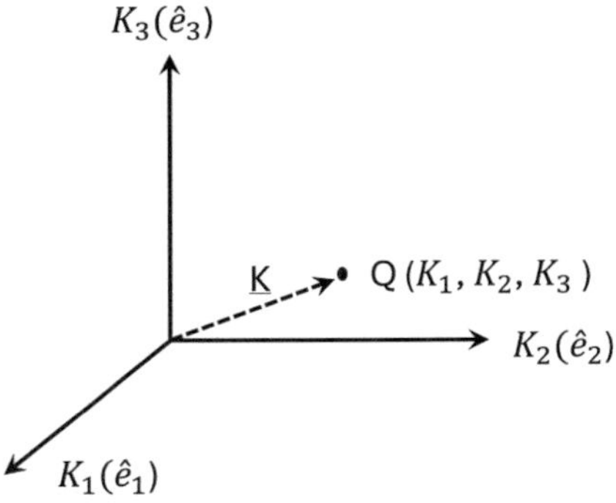

Fig. 7.2 The spectral or the Fourier space. $\underline{K}$ is an arbitrary wavenumber vector

The function $E(K)$ is also called the *energy spectrum function*.

Since the wavenumber vectors themselves are independent quantities (7.12), we can visualize a three-dimensional space wherein the Cartesian axes represent the three scalar components of an arbitrarily chosen wavenumber vector (Fig. 7.2).

$$K_1 = \frac{2\pi}{L}n_1,\ K_2 = \frac{2\pi}{L}n_2 \text{ and } K_3 = \frac{2\pi}{L}n_3 \tag{7.61}$$

Such a three-dimensional space is called the *Fourier space* or the *spectral space*. In such a space, the set of all possible wavenumber vectors, $\underline{K}^{'}$, which satisfy (7.59) are the position vectors of various locations, all of which lie within a thin shell of thickness ΔK and radius K. Thus, it follows from (7.59) and (7.60), the quantity $E(K)\Delta K$ represents the part of turbulence kinetic energy (k) which is present inside this thin shell in the Fourier space. Based on our earlier discussion in Sect. 7.1, an independently chosen value of K can be interpreted as a length-scale of the flow that exists in the cubical domain. Thus, the quantity $E(K)\Delta K$ (7.58) is often interpreted as the part of turbulence kinetic energy (k) which is associated with a length-scale $l_K = \frac{2\pi}{K}$ or with an eddy of characteristic length l_K. The spectral density function (7.58) provides us with a mathematical tool which can quantify the contribution of various length-scales to the turbulence kinetic energy per unit mass (k) that is present in the flow domain.

Further, we define the spectral density function of ϵ. In (7.52), we set

$$\phi_1 = \frac{\partial V_i^{'}}{\partial x_k} \text{ and } \phi_2 = \frac{\partial V_i^{'}}{\partial x_k} \tag{7.62}$$

This leads to

$$\left\langle \frac{\partial V_i^{'}}{\partial x_k}\frac{\partial V_i^{'}}{\partial x_k} \right\rangle_{\mathcal{V}} = \sum_{\underline{K}} \left[\mathcal{F}_K \left\{ \frac{\partial V_i^{'}}{\partial x_k} \right\} \mathcal{F}_K^* \left\{ \frac{\partial V_i^{'}}{\partial x_k} \right\} \right] \tag{7.63}$$

Using (7.39) on the RHS of (7.63) leads to

$$
\begin{aligned}
\left\langle \frac{\partial V_i'}{\partial x_k}\frac{\partial V_i'}{\partial x_k} \right\rangle_{\mathcal{V}} &= \sum_{\underline{K}} \left[\mathcal{F}_K \left\{ \frac{\partial V_i'}{\partial x_k} \right\} \mathcal{F}_K^* \left\{ \frac{\partial V_i'}{\partial x_k} \right\} \right] \\
&= \sum_{\underline{K}} \left[\left\{ i K_k \widehat{V}_i(\underline{K}) \right\} \left\{ i K_k \widehat{V}_i(\underline{K}) \right\}^* \right] \\
&= \sum_{\underline{K}} \left[-i^2 K_k \widehat{V}_i(\underline{K}) K_k \widehat{V}_i(\underline{K})^* \right] \\
&= \sum_{\underline{K}} \left[K_k K_k \widehat{V}_i(\underline{K}) \widehat{V}_i^{\,*}(\underline{K}) \right] \\
&= \sum_{\underline{K}} \left[K^2 \widehat{V}_i(\underline{K}) \widehat{V}_i^{\,*}(\underline{K}) \right]
\end{aligned}
\tag{7.64}
$$

where $K^2 = K_q K_q$ is the square of the magnitude of the wavenumber vector $\underline{K}$. Multiplying both sides by ν and using the fact that in homogeneous turbulence, volume averaging leads to an estimate of the expected value of a random quantity of interest, we are led to the following expression for the dissipation rate (ϵ)

$$
\epsilon = \nu \left\langle \frac{\partial V_i'}{\partial x_k}\frac{\partial V_i'}{\partial x_k} \right\rangle_{\mathcal{V}} = \nu \sum_{\underline{K}} \left[K^2 \widehat{V}_i(\underline{K}) \widehat{V}_i^{\,*}(\underline{K}) \right]
\tag{7.65}
$$

Accordingly, we define the spectral density function of ϵ as

$$
D(K) = \lim_{\Delta K \to 0} \nu \sum_{\underline{K}'} K'^2 \widehat{V}_i(\underline{K}') \widehat{V}_i^*(\underline{K}')
\tag{7.66}
$$

where the summation of the RHS of (7.58) is over all those wavenumber vectors $\underline{K}'$ such that

$$
K \le |\underline{K}'| < K + \Delta K
\tag{7.67}
$$

The function $D(K)$ is also the *dissipation-rate spectrum function*. It follows from the definition of E(K) that

$$
\int_0^\infty D(K) dK = \epsilon
\tag{7.68}
$$

The quantity $D(K)\Delta K$ (7.66) can be interpreted as the part of the dissipation rate (ϵ) which is associated with a length-scale $l_K = \frac{2\pi}{K}$. The spectral density function (7.66) provides us with a mathematical tool which can quantify the contribution of various length-scales to the dissipation rate of turbulence kinetic energy per unit mass (ϵ) that is present in the flow field.

7.3 Evolution of Energy and Dissipation Spectra in Decaying Turbulence: DNS-Based Observations

To examine how the spectra of turbulence kinetic energy and its dissipation rate evolve in time, we refer to the DNS database of incompressible decaying turbulence. This simulation has been performed over a cubical domain, with the initial Reynolds number based on the Taylor microscale being 40. The initial velocity field is generated such that its spectral density function is described by the function

$$E(K) = A_0 K^4 e^{(-2\,K^2/K_0^2)} \tag{7.69}$$

where A_0 and $K_0(=1)$ are constants. This initial spectral density function of k is plotted in Fig. 7.3. More details about such a simulation are available in [7]. The horizontal axis is for the quantity K (wavenumber vector magnitude). The curve of $E(K)$ shows that initially k is concentrated in a narrow neighborhood of wavenumber vectors with magnitude $K = 1$. In other words, the turbulence kinetic energy is concentrated in a narrow neighborhood of the length-scale $l_K = 2\pi$). As time progresses, the turbulent flow field evolves. We list some important observations based on the results of this DNS simulation.

1. The temporal variation in k and ϵ is shown in Fig. 7.4. In this figure, the horizontal axis represents normalized time $t^{'}$.

$$t^{'} = t/\tau \text{ where } \tau = \frac{\lambda}{V_{\text{rms}}} \tag{7.70}$$

The symbols λ and V_{rms} are defined in (7.9) and (7.8), respectively. The quantity τ is called the *eddy turnover time* of the simulation. Evidently, the turbulence kinetic energy does decay monotonically. However, the variation in ϵ is non-monotonous. ϵ first rapidly increases, reaches a peak value, and subsequently reduces.

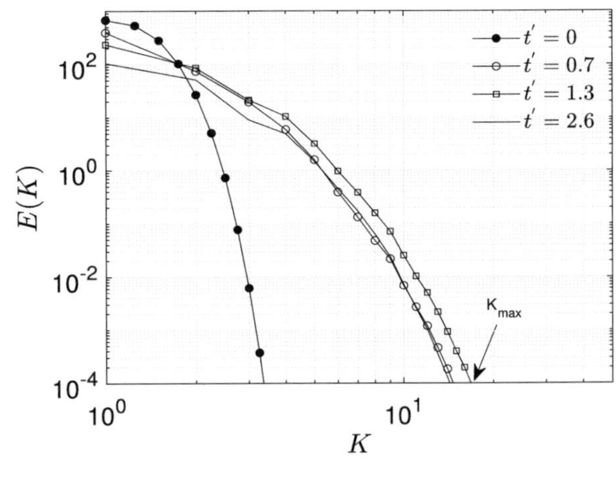

Fig. 7.3 Spectral density function of k at different time instants in a simulation of decaying turbulence

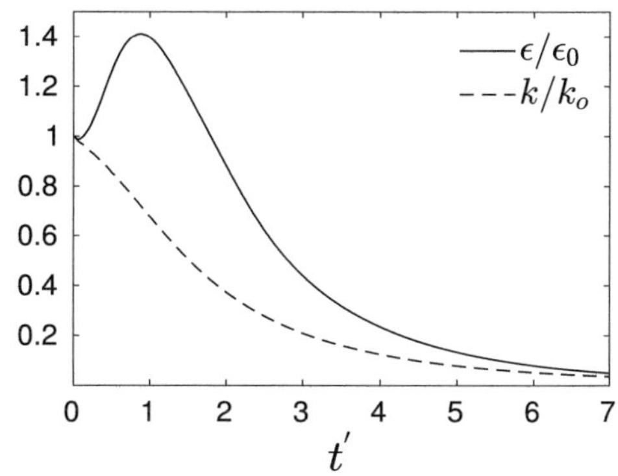

Fig. 7.4 Temporal evolution of k and ϵ in decaying turbulence. The symbols k_o and ϵ_o represent the initial values of k and ϵ

2. To understand the corresponding evolution of length-scales in the same flow field, in Fig. 7.3 we present the spectral density function $(E(K))$ of k at three more representative time instants. These normalized time instants are $t' = 0.7$ (before the peak event of ϵ in Fig. 7.4), $t' = 1.3$ (close to the peak event) and $t' = 2.6$ (after the peak event) As time progresses, it is evident that the energy spectrum spreads to larger values of K. In other words, as time progresses, smaller-length-scales are generated, and those scales also have some contribution to k. This process of the spread of turbulence kinetic energy to smaller length-scales is also called the *cascade of turbulence kinetic energy* or simply as the *energy cascade* process. These new smaller scales of motion can be related to the appearance of eddies of smaller diameters in the flow domain.
3. There seems to be an upper limit on K beyond which $E(K)$ does not spread further (assuming $E(K) < 10^{-4}$ to be of negligible significance in Fig. 7.3). We denote this value of K by $K_{\max}$. Thus,

$$k = \int_0^{\infty} E(K)dK \approx \int_0^{K_{\max}} E(K)dK \tag{7.71}$$

4. To quantify the distribution of energy over the new length-scales that have been generated in the flow field, define a new quantity f_k:

$$f_k = \frac{\int_0^{3K_0} E(K)dK}{k} \tag{7.72}$$

where K_0 is the magnitude of the wave number vector in the neighborhood of which all the initial k were concentrated. The quantity f_k represents the fraction of instantaneous turbulence kinetic energy that is associated with those length-scales (l_K) such that,

$$l_K > \frac{2\pi}{3K_0} \tag{7.73}$$

This range does represent the largest length-scales present in the flow field. In Table 7.1, we present computed values of f_k at different time instants in the same

DNS simulation. We observe that at time $t = 0$, $f_k \approx 1.0$, which simply means that all initial turbulence kinetic energy was associated with the largest length-scales. We observe that as time progresses, f_k reduces. However, evidently, at all three representative time instants (pre-occurrence of/close to/post-occurrence of the peak event) $f_k \geq 0.7$. Thus, the bulk of turbulence kinetic energy continues to be concentrated at large scales ($l_K > \frac{2\pi}{3K_0}$) during the entire decay process.

5. To quantify the distribution of dissipation rate (ϵ) over the new length-scales that have been generated in the flow field, we define a new quantity: f_ϵ

$$f_\epsilon = \frac{\int_0^{3K_0} D(K)dK}{\epsilon} \tag{7.74}$$

The quantity f_ϵ represents the fraction of instantaneous dissipation rate that is associated with those length-scales (l_K) such that,

$$l_K > \frac{2\pi}{3K_0} \tag{7.75}$$

In Table 7.1, we have included the computed values of f_ϵ, as well, from the same DNS simulation at the same three time instants where we earlier examined f_k. We observe that at time $t = 0$, f_ϵ, too, is quite high (≈ 0.9). This is so because the length-scales in the range included in (7.69) were the only length-scales present in the flow field. However, we observe that as time progresses, f_ϵ decreases rapidly and tends to reach a significantly smaller value (0.4) compared to what f_k attains at the corresponding instants. Thus, we conclude that the bulk of dissipation tends to move to newly created smaller scales in the flow field. In other words, the bulk of ϵ tends to be associated with the smaller eddies of the flow domain. This behavior is in contrast with what we observed for k, the bulk of which tends to remain associated with the largest length-scales. A simple mathematical explanation of this contrasting behavior emerges if we compare the expressions of $E(K)$ (7.58) and $D(K)$ (7.66). Since ϵ is proportional to the spatial gradients of the fluctuations, the expression of $D(K)$ has the square of the magnitude of the wavenumber vector (K^2) as an amplifying factor.

$$D(K) = 2\nu K^2 E(K) \tag{7.76}$$

As kinetic energy spreads to larger K's during the cascade process, $D(K)$ gets more amplified by large values of K^2 resulting into enhanced contribution to ϵ from large K's (small length-scales).

Table 7.1 Variation in f_k and f_ϵ at different time instants of a simulation of decaying turbulence with initial $Re_\lambda = 40$

	$t' = 0$	$t' = 0.7$	$t' = 1.3$	$t' = 2.6$
f_k	1.0	0.8	0.8	0.8
f_ϵ	0.9	0.5	0.4	0.4

6. It is observed that in simulations with higher initial Re_λ, f_ϵ tends to become smaller compared to its value in simulations with smaller initial Re_λ. Following this trend, it is expected that as Re_λ increases, a higher contribution to ϵ would come from the smallest scales in the flow field.

7.4 Explanation of Energy Cascade: Fourier Description of Navier-Stokes Equation

With the motivation to develop insights into the processes that cause the generation of new, smaller length-scales in a turbulent flow field, in this section, we wish to specifically derive and examine the evolution equation of the quantity $\widehat{V}_i(\underline{K})$. Such an equation can possibly help us understand how a particular mode of the velocity field, which has been nonexistent initially in a turbulent flow field, comes into existence at a later time.

As the first step toward deriving the evolution equation of the quantity $\widehat{V}_i(\underline{K})$, we subject the continuity (3.10) and the momentum equations (3.11) to the $\mathcal{F}_K$ operator of (7.34). Like the previous sections of this chapter, our focus is on a decaying turbulent flow field wherein $V_i = V_i^{'}$ and $\mathcal{F}_K\{V_i\} = \mathcal{F}_K\left\{V_i^{'}(\underline{K})\right\} = \widehat{V}_i(\underline{K})$. Even though all Fourier amplitudes are, in general, functions of time, for algebraic brevity, we do not include the variable t as an argument of these quantities: it is implied.

The continuity equation, when subjected to the $\mathcal{F}_K$ operator (7.34), transforms as:

$$\begin{aligned}\mathcal{F}_K\left\{\frac{\partial V_q}{\partial x_q}\right\} &= \mathcal{F}_K\{0\}\\ iK_q\widehat{V}_q(K) &= 0\\ K_q\widehat{V}_q(K) &= 0\end{aligned} \tag{7.77}$$

The final form of (7.77) is the constraint that the continuity equation (3.10) imposes on the vector $\widehat{\underline{V}}(\underline{K})$. Equation (7.77) implies that the vector $\underline{V}(\hat{\underline{K}})$ must be perpendicular to the corresponding wavenumber vector $(\underline{K})$ at all time instants.

Next, we subject the momentum equation (3.11) to the Fourier operator.

$$\mathcal{F}_K\left\{\frac{\partial V_j}{\partial t} + V_k\frac{\partial V_j}{\partial x_k}\right\} = \mathcal{F}_K\left\{-\frac{1}{\rho}\frac{\partial p}{\partial x_j} + \nu\frac{\partial^2 V_j}{\partial x_k \partial x_k}\right\} \tag{7.78}$$

Since $V_k = V_k'$ (7.55) and $\frac{\partial p}{\partial x_i} = \frac{\partial p'}{\partial x_i}$ (7.3),

$$\mathcal{F}_K\left\{\frac{\partial V_j'}{\partial t} + V_k'\frac{\partial V_j'}{\partial x_k}\right\} = \mathcal{F}_K\left\{-\frac{1}{\rho}\frac{\partial p'}{\partial x_j} + \nu\frac{\partial^2 V_j'}{\partial x_k \partial x_k}\right\} \tag{7.79}$$

Using (7.35) in (7.79) leads to

$$\underbrace{\mathcal{F}_K\left\{\frac{\partial V_j'}{\partial t}\right\}}_{I}+\underbrace{\mathcal{F}_K\left\{V_k'\frac{\partial V_j'}{\partial x_k}\right\}}_{II}=\underbrace{\mathcal{F}_K\left\{-\frac{1}{\rho}\frac{\partial p'}{\partial x_j}\right\}}_{III}+\underbrace{\mathcal{F}_K\left\{\nu\frac{\partial^2 V_j'}{\partial x_k\partial x_k}\right\}}_{IV} \tag{7.80}$$

Using (7.40) for the term involving a time derivative and (7.39) for the terms involving spatial derivatives, (7.81) is expressed as

$$\frac{d\widehat{V}_j(\underline{K})}{dt}+\widehat{G}_j(\underline{K})=-iK_j\widehat{p}(\underline{K})-\nu K^2\widehat{V}_j(\underline{K}) \tag{7.81}$$

where

$$\widehat{p}(\underline{K})=\mathcal{F}_K\left\{\frac{p'(\underline{X})}{\rho}\right\} \tag{7.82}$$

and

$$\widehat{G}_j(\underline{K})=\mathcal{F}_K\left\{G_j(\underline{X})\right\}\text{ and }G_j=V_k'\frac{\partial V_j'}{\partial x_k} \tag{7.83}$$

The exact expression of $\widehat{G}(\underline{K})$ can be obtained in terms of the Fourier amplitudes of various velocity components. Employing the continuity equation (3.10), we first express G_j in an alternate form

$$G_j(\underline{X})=V_q'\frac{\partial V_j'}{\partial x_q}=\frac{\partial(V_j'V_q')}{\partial x_q} \tag{7.84}$$

Subsequently, we subject (7.84) to the $\mathcal{F}_K$ operator.

$$\begin{aligned}
\widehat{G}_j(\underline{K})&=\mathcal{F}_K\left\{G_j(\underline{X})\right\}=\mathcal{F}_K\left\{\frac{\partial(V_j'V_q')}{\partial x_q}\right\}\\
&=iK_q\mathcal{F}_K(V_q'V_j')\\
&=iK_q\mathcal{F}_K\left\{\left(\sum_{\underline{K}'}\widehat{V}_j(\underline{K}')e^{i\underline{K}'\cdot\underline{X}}\right)\left(\sum_{\underline{K}''}\widehat{V}_q(\underline{K}'')e^{i\underline{K}''\cdot\underline{X}}\right)\right\}\\
&=iK_q\left\langle\left(\sum_{\underline{K}'}\widehat{V}_j(\underline{K}')e^{i\underline{K}'\cdot\underline{X}}\right)\left(\sum_{\underline{K}''}\widehat{V}_q(\underline{K}'')e^{i\underline{K}''\cdot\underline{X}}\right)e^{-i\underline{K}\cdot\underline{X}}\right\rangle_{\mathcal{V}}\\
&=iK_q\sum_{\underline{K}'}\sum_{\underline{K}''}\left[\widehat{V}_j(\underline{K}')\widehat{V}_q(\underline{K}'')\left\langle e^{i\underline{K}'\cdot\underline{X}}e^{i\underline{K}''\cdot\underline{X}}e^{-i\underline{K}\cdot\underline{X}}\right\rangle_{\mathcal{V}}\right]\\
&=iK_q\sum_{\underline{K}'}\sum_{\underline{K}''}\left[\widehat{V}_j(\underline{K}')\widehat{V}_q(\underline{K}'')\delta_{\underline{K}'+\underline{K}'',\,\underline{K}}\right]
\end{aligned}$$

$$= iK_q \sum_{\underline{K}'} \left[\widehat{V}_j(\underline{K}')\widehat{V}_q(\underline{K} - \underline{K}') \right] \tag{7.85}$$

Thus, the Fourier amplitude of the function $G_j(\underline{X})$ corresponding to the wavenumber vector $\underline{K}$ involves a summation of the product of Fourier amplitudes of relevant velocity components over all possible wavenumber vectors (represented by $\underline{K}'$ in the summation appearing on the RHS of 7.85).

Using (7.85), (7.81) is now expressed as

$$\frac{d\widehat{V_j}(\underline{K})}{dt} = -iK_q \sum_{\underline{K}'} \left[\widehat{V}_j(\underline{K}')\widehat{V}_q(\underline{K} - \underline{K}') \right] - iK_j\widehat{p}(\underline{K}) - \nu K^2\widehat{V_j}(\underline{K}) \tag{7.86}$$

Further, simplification of (7.86) can be achieved by finding an expression for $K_j\widehat{p}(\underline{K})$ appearing on the RHS of (7.86) in terms of $\widehat{G_j}$. This particular relationship is derived by first contracting (7.86) with the wavenumber vector ($\underline{K}$).

$$\begin{aligned} K_j\frac{d\widehat{V}_j(\underline{K})}{dt} &= -K_j iK_q \sum_{\underline{K}'} \left[\widehat{V}_j(\underline{K}')\widehat{V}_q(\underline{K} - \underline{K}') \right] \\ &\quad - iK_jK_j\widehat{p}(\underline{K}) - \nu K_jK^2\widehat{V}_j(\underline{K}), \text{ or} \\ \frac{d(K_j\widehat{V}_j(\underline{K}))}{dt} &= -K_j iK_q \sum_{\underline{K}'} \left[\widehat{V}_j(\underline{K}')\widehat{V}_q(\underline{K} - \underline{K}') \right] \\ &\quad - iK^2\widehat{p}(\underline{K}) - \nu K^2K_j\widehat{V}_j(\underline{K}) \end{aligned} \tag{7.87}$$

where $K^2 = K_jK_j$ is the square of the magnitude of the wavelength vector $\underline{K}$.

Now using the constraint (7.77), (7.87) simplifies to

$$0 = -K_j iK_q \sum_{\underline{K}'} \left[\widehat{V}_j(\underline{K}')\widehat{V}_q(\underline{K} - \underline{K}') \right] - iK^2\widehat{p}(\underline{K}) \tag{7.88}$$

Using the expression of (7.85), (7.88) is expressed in a more compact manner as

$$iK_j\widehat{p}(\underline{K}) = -K_j\frac{K_m\widehat{G}_m(\underline{K})}{K^2} \tag{7.89}$$

We can now employ (7.89) to express (7.86) in the following form

$$\frac{d\widehat{V}_j(\underline{K})}{dt} = -\left(\delta_{jm} - \frac{K_jK_m}{K^2}\right)\widehat{G}_m(\underline{K}) - \nu K^2\widehat{V}_j(\underline{K}) \tag{7.90}$$

which is the evolution equation of $\widehat{V}_j(\underline{K})$ in its most simplified form.

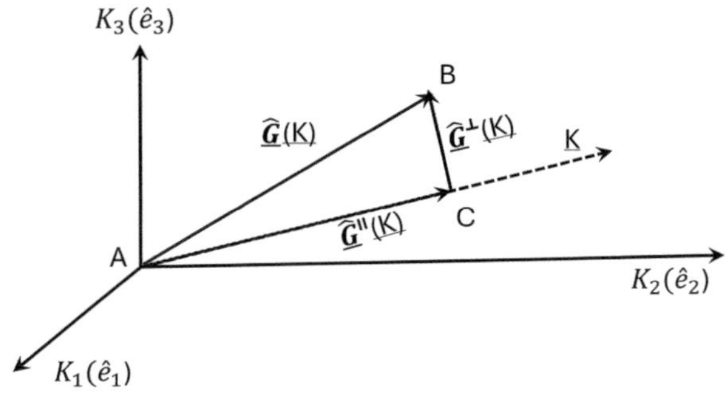

Fig. 7.5 Geometric decomposition of the vector $\underline{\hat{G}}(\underline{K})$

We can show that the first term on the RHS (7.90) is the projection of $\underline{\widehat{G}}(\underline{K})$ along the wavenumber vector $\underline{K}$. Like any other vector, we can split the vector $\underline{\widehat{G}}(\underline{K})$ into two components,

$$\underline{\widehat{G}}(\underline{K}) = \underline{\widehat{G}}^{||}(\underline{K}) + \underline{\widehat{G}}^{\perp}(\underline{K}) \tag{7.91}$$

where $\underline{\widehat{G}}^{||}(\underline{K})$ and $\underline{\widehat{G}}^{\perp}(\underline{K})$ are the projections of the vector $\underline{\widehat{G}}(\underline{K})$ along and perpendicular to the vector $\underline{K}$. Figure 7.5 we show this decomposition (7.91). In the figure, the vector $\underline{AC}$ and $\underline{AB}$ equal $\underline{\widehat{G}}^{||}(\underline{K})$ and $\underline{\widehat{G}}^{\perp}(\underline{K})$, respectively.

We can show that

$$\widehat{G}_j^{||}(\underline{K}) = \frac{K_j K_m \widehat{G}_m}{K^2} \tag{7.92}$$

and thus, according to (7.91),

$$\widehat{G}_j^{\perp}(\underline{K}) = \widehat{G}_j(\underline{K}) - \frac{K_j K_m \hat{G}_m(\underline{K})}{K^2} = \left(\delta_{jm} - \frac{K_j K_m}{K^2}\right) \widehat{G}_m(\underline{K}). \tag{7.93}$$

Thus, (7.90) can alternatively be expressed as

$$\frac{d\widehat{V}_j(\underline{K})}{dt} = \underbrace{-\widehat{G}_j^{\perp}(\widehat{K})}_{I} \underbrace{-\nu K^2 \widehat{V}_j(\underline{K})}_{II} \tag{7.94}$$

Equation (7.94) shows that $\widehat{V}_i(\underline{K})$ evolves due to the action of two processes. Process I is the one that originated from the vector $\underline{G}$ (7.83), which represents the nonlinear advection process in the physical space. The second process (II) represents a viscous process. Indeed, the viscous process equals $\widehat{V}_j(\underline{K})$ with a negative sign and a multiplication factor which is positive-definite (νK^2). Thus, the process II of (7.94) must cause a monotonic decay of the quantity $\widehat{V}_j(\underline{K})$ with time. This monotonic decaying action is eventually responsible for converting the kinetic energy associated with the wavenumber $\underline{K}$ to the internal energy of the fluid. The presence of K^2 as a multiplication factor in II suggests that the process of converting the kinetic energy associated with velocity fluctuations is magnified at wavenumber vectors which have large magnitudes. Based on this insight, we conclude the dissipation process (conversion of turbulence kinetic energy to heat) must predominantly happen at small length-scales (in modes with large values of K). This insight is in line with the observations in Table 7.1.

The viscous process of $\widehat{V}_j(\underline{K})$ (7.90) is a *local process in the Fourier space*, because it involves the Fourier amplitude of velocity corresponding to the same wavenumber vector $\underline{K}$ which appears in the quantity $\widehat{V}_j(\underline{K})$ on the LHS of (7.90). Clearly, such a local process can never distribute energy to other length-scales. Thus, this process cannot be held responsible for the generation of newer length-scales in a turbulent flow field.

In contrast, process I is *non-local in Fourier space*. This is evident by its algebraic form itself.

$$\underbrace{-\widehat{G}_j^{\perp}(\widehat{K})}_{I} = \left[\left(\delta_{mj} - \frac{K_m K_j}{K^2} \right) i K_q \sum_{\underline{K}'} \left[\widehat{V}_m(\underline{K}') \widehat{V}_q(\underline{K} - \underline{K}') \right] \right] \tag{7.95}$$

First, the process I has a summation involving Fourier amplitudes of velocity over all possible wavenumber vectors ($\underline{K}'$). Further, in every term in this summation two additional kinds of Fourier amplitudes are involved: $\widehat{V}_i(\underline{K}')$, and $\widehat{V}_i(\underline{K} - \underline{K}')$ along with the presence of the components of the wavenumber vector $\underline{K}$, itself. Owing to this involvement of three different wavenumber vectors in each term of the summation, Process I is also called a *triadic process*. Indeed, this interaction of the Fourier amplitudes at two different wavenumber vectors, $\widehat{V}_i(\underline{K}')$, and $\widehat{V}_i(\underline{K} - \underline{K}')$ contributing to the evolution of a different Fourier amplitude $\widehat{V}_i(\underline{K})$ gives this process the ability to generate newer length-scales in the flow field.

In summary, the derivation and the subsequent examination of the evolution equation of $\widehat{V}_i(\underline{K})$ (7.94) have provided us with new insights on the role of the advection and viscous processes. While the advection process tends to create newer scales of motion, the viscous process converts the kinetic energy present in the fluctuating velocity field to heat. Since this converting action of the viscous process is amplified by the square of the magnitude of the wavenumber vector, the viscous dissipative action is expected to be more prominent in those modes which have large values of K (small length-scales).

Any further mathematical analysis of various processes of (7.94) is deemed outside the scope of this book. The reader is referred to [8] and other cited works therein to have a more advanced analysis and discussion on these processes.

7.5 Kolmogorov's Hypotheses

Russian mathematician Andrey Kolmogorov (1903–1987) put forward a set of hypotheses which when interpreted using arguments involving dimensions of various statistical quantities and their orders of magnitude leads to some insightful conclusions about the cascade process in turbulent flows. These hypotheses were proposed

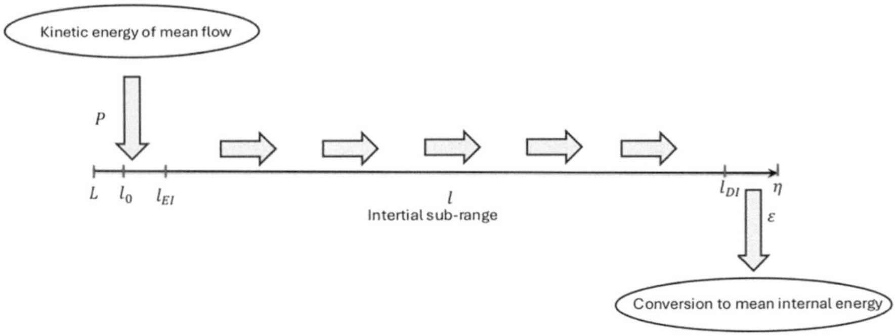

Fig. 7.6 Schematic diagram of the cascade process (Not to scale). P represents the production rate of turbulence kinetic energy

during an era when the computational fluid dynamics (CFD) tools and computing hardware were not available to perform direct numerical simulations of turbulent flow fields.

There is no mathematical proof available for these hypotheses. Nonetheless, these hypotheses have been employed as the cornerstones based on which our understanding of turbulent flows has evolved. Various cues emerging from these hypotheses have also been leveraged for turbulence modeling (some such aspects are discussed in Chap. 8). With the advent of more advanced computing hardware, several direct numerical simulations of turbulent flow fields have been performed in recent years, at increasingly high Reynolds numbers. Such numerical simulations are providing new opportunities to further examine these hypotheses.

Kolmogorov's hypotheses refer to the cascade process, about which some numerical evidence and insights have already been provided in our previous sections. Based on this background, in Fig. 7.6, we present a schematic diagram of the cascade process. This will help us in our upcoming discussion on Kolmogorov's hypotheses. In Fig. 7.6, l_o represents the characteristic length-scale of the largest eddies in the flow field. On the other hand, η represents the characteristic length-scale of the smallest eddies. The symbol L represents the characteristic length-scale of the flow domain itself. The order of magnitude of L is almost the same as the order of magnitude of l_o. We introduce two more length-scales in this diagram. The symbol l_{EI} is defined such that the bulk of k lies within the wavenumber $K < \frac{2\pi}{L_{EI}}$. The symbol l_{DI} is defined such that the bulk of ϵ lies beyond the wavenumber $K > \frac{2\pi}{L_{EI}}$. We call the range of length-scales between L and l_{DI} as the *energy-containing range*.

In our DNS simulation cases of decaying turbulence discussed earlier in this chapter (Fig. 7.4), the turbulence kinetic energy was deliberately initiated in this range of scales, and was subsequently allowed to spread to smaller scales as determined by the governing equations of motion. The production process by itself is absent in decaying turbulence (statistically homogeneous flow field, Eq. 7.6). However, in a general flow field wherein the production process of the turbulence kinetic is non-zero, turbulence kinetic energy is injected into the cascade process by the production mechanism (5.56). Mathematically, the production mechanism of the turbulence kinetic energy (5.56) is the double-dot interaction between the Reynolds stress tensor and the mean velocity gradient tensor.

Various components of the Reynolds stress tensor (R_{ij}), which are the mean values of products of fluctuating components of the velocity field (like k itself), are associated mainly with the largest eddies. Further, the mean velocity gradient scales as V_o/L, where V_o and L represent the characteristic velocity associated with the boundary conditions and the length-scale associated with the geometry of the flow domain. The production process (depicted by P in Fig. 7.6), even if it is non-zero, is predominantly a large-scale process. For such flow fields, the production rate of turbulence kinetic energy still happens in the same range of length-scales: (l_{EI}, L) as shown schematically in Fig. 7.6.

On the other end of the Fig. 7.6, the range of length-scales between l_{DI} and η is called the *dissipation range*, and the range of length-scales between the l_{EI} and l_{DI} is called the *inertial subrange*. This hierarchy of length-scales is visualized as eddies of different diameters. At sufficiently high Reynolds numbers, the process of dissipation tends to happen predominantly in the dissipation range of scales (eddies). At such high Reynolds numbers, there may not be significant dissipation happening in the inertial subrange or the energy-containing scales. Thus we draw three inferences:

1. The rate ($Jkg^{-1}s^{-1}$) at which energy is transferred across the interfacing length-scale l_{DI} (we denote this rate by the symbol T_{DI}) must be approximately the same as ϵ, itself.
2. The rate ($Jkg^{-1}s^{-1}$) at which energy is transferred across the interfacing length-scale l_{EI} (we denote this rate by the symbol T_{EI}) must also be approximately to be the same as ϵ, itself.
3. The rate ($Jkg^{-1}s^{-1}$) at which energy is transferred across any arbitrary length-scale l where $l_{EI} > l > l_{DI}$ (we denote this rate by the symbol T_l) must also be approximately the same as ϵ, itself.

The amount of kinetic energy that is dissipated per unit mass per unit time in the range of dissipative scales is sourced originally from the largest eddies and is successively transferred to smaller eddies. Finally, in the dissipative scales, this energy is converted to heat. In Fig. 7.6, the arrows depict the direction of flow of turbulence kinetic energy (k) across various scales of motion.

Using arguments based on dimensionality, we estimate the order of magnitude of the energy transfer rate from the largest eddies ($Jkg^{-1}s^{-1}$) to the smaller ones is $\frac{u_o^3}{l_o}$. The symbols u_o and l_o denote the characteristic velocity and length-scale of the largest eddies in the flow field. Based on this estimate, we summarize the cascade process as described in the previous paragraph as,

$$\frac{u_o^3}{l_o} \sim T_{EI} \sim T_l \sim T_{DI} \sim \epsilon \tag{7.96}$$

where the symbol $\sim$ means the equality in the orders of magnitude of the quantities on the left-hand and right-hand sides of the symbol.

7.5.1 Kolmogorov's First Similarity Hypothesis

Kolmogorov's first similarity hypothesis states *"In every turbulent flow at sufficiently high Reynolds number, the statistics of small-scale-motion have a universal form determined by ν and ϵ"* [6]. Here ν and ϵ denote the coefficient of kinematic viscosity of the fluid and the rate of dissipation of turbulence kinetic energy per unit mass (5.58), The Reynolds number (Re) in the context is one based on the characteristic length (l_o) and velocity scale (u_o) of the largest eddies

$$Re = \frac{u_o l_o}{\nu} \tag{7.97}$$

This hypothesis can be employed to estimate the orders of magnitude of the characteristic length, time and velocity scales of the smallest eddies in a turbulent flow field. We use the symbols η, u_η and τ_η to denote these quantities, respectively. Based on Kolmogorov's first similarity hypothesis, we infer

$$\eta \sim \epsilon^a \nu^b \tag{7.98}$$

$$u_\eta \sim \epsilon^c \nu^d \tag{7.99}$$

$$\tau_\eta \sim \epsilon^e \nu^f \tag{7.100}$$

The symbols a, b, c, d, e and f are constants to be determined.

Using the dimensions of η (m), u_η (ms^{-1}) and τ_η (s) on one hand and those of ϵ $(Jkg^{-1}s^{-1})$ and ν (m^2s^{-1}) on the other hand, leads to a system of six linear algebraic equations in six unknowns coefficients which are appearing as exponents in (7.98–7.100). Solving this set of equations leads to

$$(a, b, c, d, e, f, g) = \left(-\frac{1}{4}, \frac{3}{4}, \frac{1}{4}, \frac{1}{4}, -\frac{1}{2}, \frac{1}{2}\right) \tag{7.101}$$

Thus, Kolmogorov's first similarity hypothesis leads to the following characteristic measures of the smallest eddies in a turbulent flow field.

$$\eta \sim \left(\frac{\nu^3}{\epsilon}\right)^{1/4} \tag{7.102}$$

$$u_\eta \sim (\epsilon\nu)^{1/4} \tag{7.103}$$

$$\tau_\eta \sim \left(\frac{\nu}{\epsilon}\right)^{1/2} \tag{7.104}$$

If we define a Reynolds number specifically in the context of the smallest eddies, it can be verified that such a Reynolds number (Re_η) has its order of magnitude as unity.

$$Re_\eta = \frac{u_\eta \eta}{\nu} \sim 1 \tag{7.105}$$

The quantities η, u_η and τ_η are the *Kolmogorov length-scale*, the *Kolmogorov velocity scale* and the *Kolmogorov time scale*, respectively.

The order of magnitude estimates arrived in (7.102)–(7.104) can be further employed to find estimates of the ratios of the characteristic features of the largest to the smallest eddies in a turbulent flow field. Using the estimate included in (7.96), we first express the order of magnitude of ϵ as

$$\epsilon \sim \frac{u_o^3}{l_o} \tag{7.106}$$

Next using this estimate of ϵ along with the estimates of the Kolmogorov length, velocity, and time scales (7.102, 7.103, and 7.104), leads to the following relationships

$$\frac{\eta}{l_o} \sim Re^{-3/4} \tag{7.107}$$

$$\frac{u_\eta}{u_o} \sim Re^{-1/4} \tag{7.108}$$

$$\frac{\tau_\eta}{\tau_o} \sim Re^{-1/2} \tag{7.109}$$

where l_o, u_o, and τ_o are the characteristic length, velocity and times scales of the largest eddies, and

$$\tau_o = \frac{l_o}{u_o}. \tag{7.110}$$

These estimates (7.107–7.109) clearly show at a higher Re, the disparity between the largest to the smallest length-scales, the disparity between the largest to the smallest velocity scales, and the disparity between the largest to the smallest time scales increase exponentially. Indeed, these are the estimates based on which we introduced the reader to the essential nature of turbulence earlier in Chap. 3 (3.6 and 3.5).

7.5.2 Kolmogorov's Second Similarity Hypothesis

Kolmogorov's second similarity hypothesis states *"In every turbulent flow at sufficiently high Reynolds number, the statistics of the motion of length-scale l in the range $l_o \gg l \gg \eta$ have a universal form that is uniquely determined by ϵ and is independent of ν"* [6]. Here l_o is the characteristic length-scale of the largest eddies existing in the turbulent flow field, and η represents the Kolmogorov length-scale.

This hypothesis can be used to estimate $u(l)$ and $\tau(l)$, where l represents a length-scale chosen independently in the range $l_o \gg l \gg \eta$ and the symbols $u(l)$ and $\tau(l)$

represent the characteristic velocity and time scales of the eddies having their characteristic length-scales as l. Following the second similarity hypothesis, we have

$$u(l) \sim \epsilon^a l^b \tag{7.111}$$

$$\tau(l) \sim \epsilon^c l^d \tag{7.112}$$

Using the dimensions of $u(l)$ (ms^{-1}) and τ_l (s) on one hand and those of ϵ $(Jkg^{-1}s^{-1})$ and l (m) on the other hand, leads to a system of four linear algebraic equations in four unknowns coefficients, which are appearing as exponents in Eqs. (7.111–7.112). Solving this set of equations leads to

$$(a, b, c, d) = \left(\frac{1}{3}, \frac{1}{3}, -\frac{1}{3}, \frac{2}{3}\right) \tag{7.113}$$

Thus, Kolmogorov's second similarity hypothesis leads to the following estimates for the eddies with their characteristic length-scales (l) lying in the range $l_o \gg l \gg \eta$.

$$u(l) \sim (\epsilon l)^{1/3} \tag{7.114}$$

$$\tau(l) \sim \left(\frac{l^2}{\epsilon}\right)^{1/3}. \tag{7.115}$$

Using (7.106), Eqs. (7.114) and (7.115) can alternatively be expressed in terms of the characteristic velocity (u_o) and the characteristic time scale (τ_o) of the largest eddies

$$u(l) \sim u_o \left(\frac{l}{l_o}\right)^{1/3} \tag{7.116}$$

$$\tau(l) \sim \tau_o \left(\frac{l}{l_o}\right)^{2/3} \tag{7.117}$$

Since $l/l_o < 1$ (7.116) and (7.117) indicate that smaller eddies have smaller characteristic time scales and smaller characteristic velocity scales compared to those of the larger eddies. The reduction in the characteristic time means that smaller eddies tend to rotate faster (higher angular velocities) as compared to the larger eddies. However, their characteristic tangential velocity, represented by $u(l)$, is smaller than that of the larger eddies.

Kolmogorov's second similarity hypothesis can be leveraged to estimate the form of the spectral density function of k, as well, in the range $l_o \gg l \gg \eta$. Following the statement of the hypothesis, we make the following conjecture

$$E(K) \sim \epsilon^a l^b \sim \epsilon^a \left(\frac{2\pi}{K}\right)^b \tag{7.118}$$

We ignore the factor 2π in (7.120), and instead concentrate on seeking the values of the exponents, a and b. The SI units of the function $E(K)$ are $Jmkg^{-1}s^{-1}$. Matching the dimensions of the two sides of (7.118) leads to a system of two linear equations in two unknown coefficients a and b. The solution of this system of equations is

$$(a, b) = \left(\frac{2}{3}, \frac{-5}{3}\right) \tag{7.119}$$

Thus, (7.118) is expressed as

$$E(K) \sim \epsilon^{2/3} K^{-5/3} \tag{7.120}$$

Thus, a plot of $lnE(K)$ versus lnK must be a straight line with its slope being $-5/3$. A straightforward implication of (7.120) is

$$D(K) = 2\nu K^2 E(K) \sim 2\nu K^2 \epsilon^{2/3} K^{-5/3} \sim 2\nu \epsilon^{2/3} K^{1/3} \tag{7.121}$$

Thus, it is expected that at a high Reynolds number and in the range $l_o \gg l \gg \eta$, the slope of $lnD(K)$ versus lnK must be $1/3$.

7.5.3 Kolmogorov's Hypothesis of Local Isotropy

Kolmogorov's hypothesis of local isotropy states that *"at sufficiently high Reynolds number, the small-scale turbulent motions (at $l \ll l_o$) are statistically isotropic"* [6]. One plausible interpretation of this hypothesis is that the statistics of the orientational tendencies of the smallest eddies in a turbulent flow field are oblivious to the boundary conditions of the turbulent flow field. While the largest eddies in a turbulent flow field are oriented following the geometric constraints of the flow domain, at smaller scales, these constraints become less important, and consequently, the smallest eddies tend to lose any preferred orientational characteristics.

8 Turbulence Modeling

Turbulence models by definition are additional approximate equations so that the RANS equation set (5.10) can be mathematically closed. Every turbulence model involves some uncertainty. It can not be an exact representation of the flow physics. Notwithstanding this realization, the computational fluid dynamics (CFD) community does need turbulence models. At the same time, any arbitrary relationship between the secondary and the primary unknowns cannot be accepted by the community as a viable turbulence model. There are, after all, features based on which a turbulence model can be deemed acceptable.

1. A turbulence model must predict the primary RANS variables ($\langle V_i \rangle$ and $\langle p \rangle$) with an acceptable level of accuracy in, at least, a few types of flow fields.
2. A turbulence model is expected to provide reasonably accurate predictions for a wider variety of flow fields.
3. A turbulence model must be proposed based on some essential physics of the flow field.
4. A turbulence model is expected not to be algebraically too complex.
5. A turbulence model is expected not to introduce any undesirable numerical issues to the overall procedure seeking a solution of the RANS equation set (5.10).
6. A turbulence model is expected not to add too much additional computational overhead to the overall numerical scheme employed for solving the RANS equation set (5.10).

In the century-long history of turbulence research, several turbulence models have been proposed, and employed for performing CFD simulations. Our goal in this chapter is not to build an exhaustive compendium of all these models. However, in the rest of the chapter, we focus on a particular category of turbulence models called the *eddy viscosity closure models*.

S. S. Sinha, *Fundamentals of Turbulence and Its Modeling*,
https://doi.org/10.1007/978-3-031-94016-3_8

8.1 Eddy Viscosity Closure

The eddy viscosity closure (also called the *turbulent viscosity closure*) is inspired by the Stokes constitutive equation for the viscous stress tensor in a Newtonian fluid. We recall that in a viscous fluid, the instantaneous viscous stress tensor (τ_{ij}) is the anisotropic part of the instantaneous local stress tensor (σ_{ij})

$$\tau_{ij} = \sigma_{ij} - \sigma_{kk}\frac{\delta_{ij}}{3} \tag{8.1}$$

where $\sigma_{kk} = -p$ (p being the local pressure value). The Stokes constitutive relationship for τ_{ij} in a Newtonian fluid is (5.18)

$$\tau_{ij} = 2\mu S_{ij} \tag{8.2}$$

where S_{ij} represents the (ij)th component of the instantaneous strain-rate tensor, and μ is a scalar which is called the coefficient of dynamic viscosity.

The essential assumption made by this model is that the tensor $\underline{\tau}$ is *aligned* perfectly with the instantaneous strain-rate tensor. In the context of a symmetric second-order tensor, the word "aligns" means that the principal coordinate system of the two tensors in context is identical. The Stokes hypothesis is a semi-empirical proposition. However, it is observed to be true for several common fluids like water and air. As temperature varies, the scalar μ may undergo some variations in Newtonian fluids. However, μ still does not any have dependence on the instantaneous velocity gradient field.

The kinetic theory of gases postulates that the instantaneous velocity of a gas molecule within a fluid particle can be expressed as [9]

$${}^{n}\underline{U} = \underline{V} + {}^{n}\underline{C} \tag{8.3}$$

where ${}^{n}\underline{U}$ represents the velocity of the nth molecule residing inside a fluid particle. On the continuum scale, a fluid particle is indeed a point mass. However, on the absolute scale, it does have non-zero dimensions. In Fig. 8.1 the shown cube represents a magnified fluid particle. Various dots represent molecules which are currently inside the fluid particle. At this instant, the centre of mass of the fluid particle is located at point Q with the position vector $\underline{X}$. The symbol $\underline{V}$ represents the velocity vector of the centre of mass of the shown fluid particle. In continuum description, this is called the velocity of the fluid particle itself. The symbol ${}^{n}\underline{C}$ represents the velocity of the nth molecule relative to the centre of mass of the fluid particle. The velocity vector ${}^{n}\underline{C}$ is called the *peculiar velocity* of the nth molecule. Following the assumptions of the kinetic theory of gases, the peculiar velocity vector is deemed to be random in nature, whereas $\underline{V}(\underline{X})$, the velocity of the fluid particle itself is deemed to be a deterministic quantity. It can be shown that the instantaneous stress tensor ($\underline{\sigma}$) arising in a continuum flow field of a gaseous medium is the ensemble-average of the moment $c_i c_j$. Thus, it is the anisotropic part of the ensemble average, of $c_i c_j$ which is eventually modelled by the Stokes constitutive relationship (8.2).

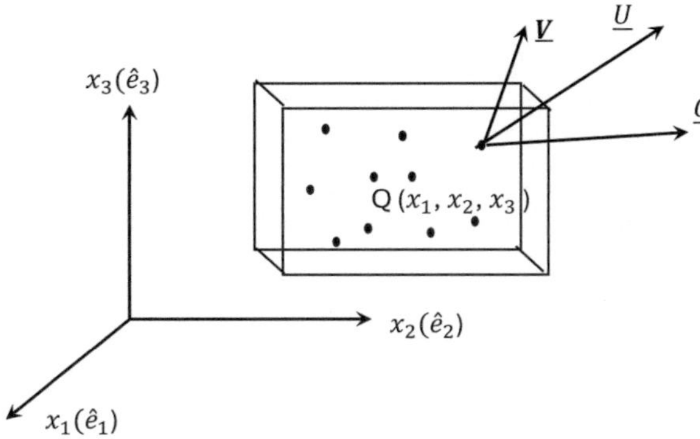

Fig. 8.1 The instantaneous, the center-of-mass and the peculiar velocity vectors of the *n*th molecule, which is currently inside the fluid particle

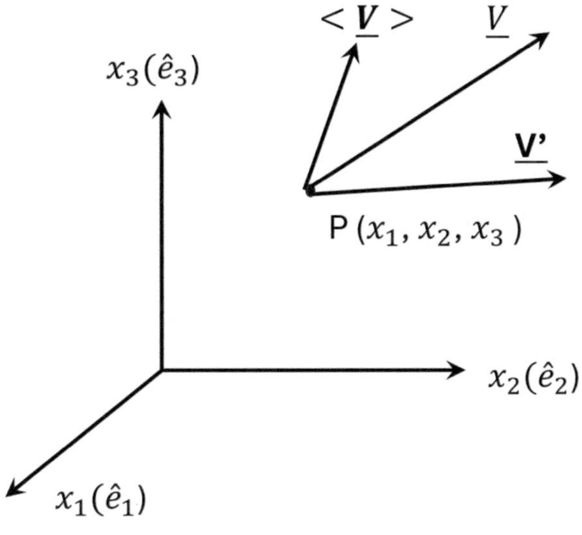

Fig. 8.2 The instantaneous, the mean and the fluctuating velocity vectors of a fluid particle in a turbulent flow field

Further, the kinetic theory of gases shows that the coefficient of dynamic viscosity equals

$$\mu = D\rho V_{\text{molecular}} l_{\text{molecular}} \tag{8.4}$$

where D is a dimensionless constant, ρ is the gas density, $V_{\text{molecular}}$ is the root-mean-square of the peculiar velocity of all the molecules that are currently inside the fluid particle of Fig. 8.1. The symbol $l_{\text{molecular}}$ represents the *mean free path* of the molecules. The quantities $V_{\text{molecular}}$ and $l_{\text{molecular}}$ can be viewed as the characteristic velocity and length scales of a transport mechanism that arises due to the random nature of the peculiar velocity of molecules.

Now let us refer to Fig. 8.2, which shows a fluid particle in a turbulent flow field. The symbols $\underline{V}$, $\langle \underline{V} \rangle$ and $\underline{V}'$ represent the instantaneous, the mean and the fluctuating velocity vector of the fluid particle which is currently located at $\underline{\mathrm{X}}$.

In the mean description of the turbulent flow field, the fluctuating velocity vector ($\underline{V}'$) of a fluid particle is deemed to be of random nature, whereas $\langle \underline{V} \rangle$, the mean velocity of the fluid particle is a deterministic quantity. Further, we have already shown that the Reynolds stress tensor component arising in the mean description of a turbulent flow field is an ensemble average of the moment $V_i^{'} V_j^{'}$.

The descriptions of quantities presented in Figs. 8.1 and 8.2 have some apparent similarities.

1. In both the figures, a pertinent velocity vector is decomposed into a deterministic part and a random part.
2. In both the scenarios, there is an unclosed tensor that arises due to the ensemble average (or mean) of a second-order moment of relevant random velocity vectors.

The eddy viscosity closure draws inspiration from these apparent similarities in Figs. 8.1 and 8.2, and takes a bold step to propose a closure for the anisotropic part of the Reynolds stress tensor ($\underline{R}^{\text{anisotropic}}$) in accordance with the Stokes constitutive relationship (8.2).

The anisotropic part of the Reynolds stress tensor ($\underline{R}^{\text{anisotropic}}$) is defined (1.17) as

$$R_{ij}^{\text{anisotropic}} = R_{ij} - R_{kk}\frac{\delta_{ij}}{3} = R_{ij} + 2\rho k\frac{\delta_{ij}}{3}. \tag{8.5}$$

Using (8.2) as the basis of an analogy, the eddy viscosity closure *assumes* that the anisotropic part of the Reynolds stress tensor is *always* perfectly aligned with the mean strain-rate tensor $\underline{S}$.

$$R_{ij}^{\text{anisotropic}} = 2\mu_T \left\langle S_{ij} \right\rangle \tag{8.6}$$

where $\left\langle S_{ij} \right\rangle$ is the (ij)th component of the mean strain-rate tensor ($\langle S \rangle$). The symbol μ_T represents a new scalar. The quantity μ_T is called the *coefficient of dynamic eddy viscosity* or the *coefficient of dynamic turbulent viscosity*. Equation (8.6) is a set of only five independent algebraic equations.

Further, the next hint towards achieving the full closure of RANS equations while using the eddy viscosity closure can be extracted by extending the underlying physical meaning of μ in the Stokes constitutive relationship. Like μ, which depends on the characteristic velocity and characteristic length scale of the random transporting action of the peculiar velocity of molecules (8.4), μ_T in (8.6) is modelled in the following form

$$\mu_T = D^{'} \rho V_{\text{fluctuation}} l_{\text{fluctuation}} \tag{8.7}$$

where $D^{'}$ is a dimensionless constant, ρ is the density of the fluid medium and $V_{\text{fluctuation}}$, and $l_{\text{fluctuation}}$ represent the characteristic velocity and a characteristic length scale associated with the fluctuating flow field. These characteristic values can be interpreted as the characteristic velocity scale and the characteristic length scale of the eddies, which carry the bulk of turbulence kinetic energy with them. Thus, $V_{\text{fluctation}}$ and $l_{\text{fluctation}}$ are modelled as the characteristic velocity scale and characteristic length scale of the largest eddies present in the turbulent flow field. With this interpretation, we realize that while μ characterizes the action of momentum transport caused by the random molecular motion (which is present in all instantaneous flow fields), μ_T characterizes the action of momentum transport caused by the fluctuating motion in the RANS description of a turbulent flow field. Since the fluctuating flow field is associated with eddying motions, μ_T is called the coefficient of dynamic eddy viscosity.

Even though the analogy with the molecular motion and the Stokes hypothesis has provided us with some bases to propose (8.6) and (8.7), we realize that we have now introduced two new scalar unknowns in our modeling procedure: $V_{\text{fluctation}}$ and $l_{\text{fluctation}}$. Even though there are physical interpretations of these quantities, they are unknowns, nonetheless. In general, we expect these quantities to be dependent on both space and time.

To make any further progress towards full closure, we will have to add additional equations describing the variations of these quantities. Indeed, over the past century, several propositions have been made with their own strength and shortcomings, leading to different closure models. In Sects. 8.3 and 8.4 of this chapter, we discuss two such popular closure strategies. While the additional closure equations will add their own modeling assumptions to the closed set of governing equations, the use of (8.6) inevitably entails the key assumptions that (a) the anisotropic part of the Reynolds stress tensor is aligned perfectly with the mean strain-rate tensor, and (b) the coefficient of eddy viscosity is a flow variable, and it is interpreted to have dependence on the characteristic velocity and length scales of the largest eddies in a turbulent flow field.

The Stokes constitutive equation (8.2) is verifiably quite accurate for Newtonian fluids. However, just because the eddy viscosity closure is mathematically analogous to the otherwise accurate Stokes relationship, it does not guarantee that the eddy viscosity closure will also be equally accurate in all types of flow fields. Even though many popular turbulence closure models have been developed based on the eddy viscosity closure paradigm, there exist certain types of flow fields wherein the assumption of the perfect alignment between the anisotropic Reynolds stress tensor and the mean strain-rate tensor fails over extended regions in the flow field. The reader is referred to [10] for further details on such failures of the model that employ the eddy viscosity closure.

Notwithstanding such failures in some specific kinds of flow fields, there is still a reasonably wide variety of engineering flow fields where the models based on the eddy viscosity closure have been used frequently with considerable success in predicting the mean flow variables. Owing to such success and popularity, in the rest of this chapter, we perform case studies on two such popular closure models: the k-ϵ model and the k-ω model.

8.2 Modeling the Unclosed Terms: A Broad Outline

These case studies are included here to help the reader gain deeper insight into the rationale followed by the respective turbulence modellers in getting the final forms of the $k - \epsilon$ and the $k - \omega$ models. Even though there is no community-wide, fully-accepted general procedure or a template with which we can explain the rationale behind every turbulence model that exists today, nonetheless, we identify a list of common steps that the developers of these two models have followed.

1. *Step I:* Some essential, relevant underlying physics of the mathematically unclosed term is invoked. The modeller's awareness and personal insights about the relevant physics of turbulent flows, as revealed in various experiment-based or direct-numerical simulation-based studies of even simple flow fields like the flat plate boundary layer flow or the decaying turbulent flow field may prove to be useful in this step of turbulence modeling.
2. *Step II:* The exact unclosed mathematical form of the to-be-modelled term is discarded and some dimensionality-based arguments to construct its alternate mathematical form that can possibly mimic that identified essential physics in Step I are proposed. Here, it must be ensured that this alternate mathematical form is expressed in terms of the quantities for which the modeller already has, or for which the modeller is planning to have additional governing equations in the final set of closed equations. Further, in this step, the modeller may also introduce some dimensionless constants of proportionality (also called the *closure coefficients* of the model) to have superior control over the performance of the closure model.
3. *Step III* Available experimental databases or direct numerical simulation databases of simple flow fields are employed to ascertain the best possible values of the closure coefficients of the model, which have been introduced in the previous step. This provides leverage to the modeller to have some fine-tuning on the performance of the model. It is expected that the modeller himself/herself prescribes the most optimal set of these closure coefficients, which can be employed for performing RANS simulation for a wide variety of flow fields. *These closure coefficients are not expected to be re-adjusted on the flow-to-flow basis or based merely on the whim and convenience of the end-user.*

Even though (a) the identification of the essential physics of the unclosed term and (b) the determination of the dimensionless constants are often done by modellers using an available experimental database or an available DNS database of some simple flow fields, the final form of the model thus proposed, are indeed expected to be employed in simulating more complex flows of engineering interest. This practice that a model which has been proposed based on our observations of some simple flow fields is then employed for more complex flow fields does lead to some modeling uncertainties. Nonetheless, in the absence of any better option, such models may still be considered acceptable as long as the performance of the model is found to be satisfactory by the CFD community in simulating their complex flows of interest.

8.3 The k-ω Model: A Case Study on Turbulence Modeling

The description of the k-ω model presented in this section is based on the work of Wilcox [11]. To model μ_T, the k-ω model introduces some variations on the RHS of (8.7). Instead of using $l_{\text{fluctuation}}$ in the expression, it employs another quantity with which we can characterize the largest eddies. This quantity is represented by the symbol ω, which represents the characteristic angular velocity of the largest eddies (the SI unit being s^{-1})

$$\omega \sim \frac{V_{\text{fluctuation}}}{l_{\text{fluctation}}}, \tag{8.8}$$

leading to the following form of μ_T (8.7)

$$\mu_T = \rho \frac{V^2_{\text{fluctuation}}}{\omega} \tag{8.9}$$

where D' has been dropped (modeller's discretion).

The k-ω model further assumes that $V_{\text{fluctation}}$ can be estimated using the local instantaneous value of k in the flow field,

$$V_{\text{fluctation}} = k^{1/2}, \tag{8.10}$$

leading to the following form of μ_T

$$\mu_T = \frac{\rho k}{\omega} \tag{8.11}$$

Thus, the two unclosed unknowns in our hands now are k and ω. In general, these quantities are expected to change with space and time. The k-ω model proposes to add (to the existing set of four partial differential equations of 5.10), two additional partial differential equations describing the evolution of k and ω in the flow field.

We have earlier derived the transport equation of k (5.56). Even though this equation is exact, it is mathematically unclosed. The specific unclosed terms are highlighted here as A, B and C.

$$\begin{aligned}\frac{\partial k}{\partial t} + \left\langle V_q \right\rangle \frac{\partial k}{\partial x_q} &= \frac{R_{ij}}{\rho}\frac{\partial \langle V_i \rangle}{\partial x_j} + \nu \frac{\partial^2 k}{\partial x_q \partial x_q} \\ &- \underbrace{\epsilon}_{A} - \underbrace{\frac{\partial}{\partial x_q}\left\langle V_q^{'} \frac{V_i^{'} V_i^{'}}{2}\right\rangle}_{B} + \underbrace{-\frac{1}{\rho}\frac{\partial \left\langle p^{'} V_i^{'}\right\rangle}{\partial x_i}}_{C}\end{aligned} \tag{8.12}$$

where $\epsilon = \nu \left\langle \frac{\partial V_i^{'}}{\partial x_q}\frac{\partial V_i^{'}}{\partial x_q}\right\rangle$.

To model ϵ, the relevant essential physics is that ϵ also equals the rate of energy transfer from the largest eddies to the smaller ones (T_{EI}, Fig. 7.6, Eq. 7.96). The k-ω model estimates ϵ in line with (7.96), as

$$\epsilon \sim \frac{V_{\text{fluctuation}}^3}{l_{\text{fluctuation}}} \tag{8.13}$$

where $l_{\text{fluctuation}}$ is the characteristic length scale of the largest eddies present in the flow field.

Using (8.8) and (8.10), in (8.9) we arrive at the following modeled equation for ϵ

$$\epsilon = \beta^* k\omega \tag{8.14}$$

where β^* is a dimensionless constant, which remains to be specified.

To model Term B in (8.12), we invoke what is generally called the *gradient diffusion hypothesis*. We first explain this hypothesis in detail in the context of an instantaneous flow field and subsequently leverage it to model Term B, specifically.

If $\phi(\underline{X}, t)$ represents an entity (such as energy, species mass, momentum) per unit mass of the background fluid, then, as per the gradient diffusion hypothesis, the amount (${}^{\phi}Q_i$) of that quantity transported per unit time per unit area across an imaginary surface having unit normal along the Cartesian unit vector $\hat{e}_i$ is assumed to be proportional to the gradient of ϕ along $\hat{e}_i$.

$${}^{\phi}Q_i \approx -\Gamma_\phi \left(\frac{\partial \phi}{\partial x_i}\right) \tag{8.15}$$

where Γ_ϕ is a scalar, and is called the *coefficient of diffusion* of ϕ in that fluid medium. The quantity ${}^{\phi}Q_i$ is called the ith component of the *flux vector*, ${}^{\phi}\underline{Q}$ of the entity in context. The negative sign on the RHS of (8.15) implies that the transport due to the gradient diffusion hypothesis is directed opposite to the gradient to ϕ. Thus, the gradient diffusion hypothesis implies that the entity in context is transported from a region of higher concentration of ϕ to a region of lower concentration of ϕ. If with the course of time, the distribution of $\phi(\underline{X}, t)$ becomes uniform in space, then, according to the gradient diffusion hypothesis, the flux of the entity (${}^{\phi}\underline{Q}$) must vanish.

In an instantaneous flow field, the coefficient of diffusion, like μ in (8.4), is the ensembled effect of the transporting action of the random peculiar velocity of molecules. Such a transport process is called the *molecular diffusion process*. *Fick's law of mass diffusion* and the *Fourier law of heat conduction* are common examples of the constitutive equations which employ the gradient diffusion hypothesis in the context of instantaneous flow fields.

We earlier discussed in Chap. 5 that Term B on the RHS of (8.12) can be expressed as the gradient of a flux term, Q_m, which represents the expected value of the flux of the fluctuating kinetic ($V_i^{'} V_i^{'}/2$) energy caused by the fluctuating velocity vector component $V_m^{'}$

$$\underbrace{\frac{\partial}{\partial x_q}\left\langle V_q^{'} \frac{V_i^{'} V_i^{'}}{2} \right\rangle}_{B} = \frac{\partial Q_m}{\partial x_m} \tag{8.16}$$

where

$$Q_m = \left\langle V_m^{'} \frac{V_i^{'} V_i^{'}}{2} \right\rangle \tag{8.17}$$

This insight that Term B is the gradient of a flux is deemed as the relevant essential physics for modeling this term. Leveraging the gradient diffusion hypothesis (8.15), we replace the exact but the unclosed mathematical form of Term B, with the following modelled form

$$Q_m = \left\langle V_m^{'} \frac{V_i^{'} V_i^{'}}{2} \right\rangle = \left\langle -\Gamma_\phi \frac{\partial}{\partial x_m} \left(\frac{V_i^{'} V_i^{'}}{2} \right) \right\rangle \tag{8.18}$$

where Γ_ϕ is a scalar that is supposed to characterize the random transporting action associated with the fluctuating velocity field, or in other words, the random transporting action associated with the largest eddies in the flow field. Since we have already chosen to employ the quantities k and ω to have the kinematic characterization of the largest eddies in the flow field (8.10 and 8.8), the k-ω model proposes to have Γ_ϕ expressed in terms of k and ω

$$\Gamma_\phi = \sigma^* k^a \omega^b \tag{8.19}$$

where σ^* is a dimensionless constant, which remains to be specified. The symbols a and b are constant coefficients. Equation (8.18) suggests that the SI units of Γ_ϕ must be $m^2 s^{-1}$. Imposing this constraint on the two sides of (8.19), we find $a = 1$ and $b = -1$. Thus, (8.18) is modified to the following form.

$$\begin{aligned} Q_m &= \left\langle V_m^{'} \frac{V_i^{'} V_i^{'}}{2} \right\rangle = \left\langle -\frac{\sigma^* k}{\omega} \frac{\partial}{\partial x_q} \left(\frac{V_i^{'} V_i^{'}}{2} \right) \right\rangle = -\frac{\sigma^* k}{\omega} \frac{\partial}{\partial x_q} \left\langle \frac{V_i^{'} V_i^{'}}{2} \right\rangle \\ &= -\frac{\sigma^* k}{\omega} \frac{\partial k}{\partial x_q} \end{aligned} \tag{8.20}$$

With (8.20), finally, Term B is assigned the following modelled form.

$$\underbrace{\frac{\partial}{\partial x_q}\left\langle V_q^{'} \frac{V_i^{'} V_i^{'}}{2} \right\rangle}_{B} = -\frac{\partial}{\partial x_q} \left(\frac{\sigma^* k}{\omega} \frac{\partial k}{\partial x_q} \right) \tag{8.21}$$

Term C is the next unclosed term in (8.12). Even though the underlying physics of this mathematical expression is known (highlighted in Chap. 5), the main modeling challenge posed by this term is the involvement of pressure fluctuations in it modeling such pressure-based physics in terms of the kinematic characteristics of the largest eddies (k and ω) is not straightforward. Thus, it is decided to drop this term altogether. In summary, the modelled form of the transport equation of k for the k-ω closure is

$$\begin{aligned}\frac{\partial k}{\partial t}+\langle V_q\rangle\frac{\partial k}{\partial x_q} &= \frac{R_{ij}}{\rho}\frac{\partial\langle V_i\rangle}{\partial x_j}+\nu\frac{\partial^2 k}{\partial x_q\partial x_q}-\beta^* k\omega+\frac{\partial}{\partial x_q}\left(\frac{\sigma^* k}{\omega}\frac{\partial k}{\partial x_q}\right)\\ &= \underbrace{\frac{R_{ij}}{\rho}\frac{\partial\langle V_i\rangle}{\partial x_j}}_{I}-\underbrace{\beta^* k\omega}_{II}+\underbrace{\frac{\partial}{\partial x_q}\left[\nu\frac{\partial k}{\partial x_q}\right]}_{III}+\underbrace{\frac{\partial}{\partial x_q}\left[\sigma^*\nu_T\frac{\partial k}{\partial x_q}\right]}_{IV}\end{aligned} \tag{8.22}$$

where we have a new symbol ν_T (which is called the *coefficient of kinematic eddy viscosity*).

$$\nu_T=\frac{k}{\omega}=\frac{\mu_T}{\rho}. \tag{8.23}$$

Term B is also called the eddy diffusion process in the governing equation of k. Equation (8.22) is symbolically expressed as

$$\frac{\partial k}{\partial t}+\langle V_q\rangle\frac{\partial k}{\partial x_q}={}^kP_I-{}^kP_{II}+{}^kP_{III}+{}^kP_{IV} \tag{8.24}$$

where the four individual processes on the RHS of this equation are identified as

$$\text{Production process }{}^kP_I=\underbrace{\frac{R_{ij}}{\rho}\frac{\partial\langle V_i\rangle}{\partial x_j}}_{I}$$

$$\text{Dissipation process }{}^kP_{II}\;:\;\underbrace{\beta^* k\omega}_{II}$$

$$\text{Molecular diffusion process }{}^kP_{III}\;:\;\underbrace{\frac{\partial}{\partial x_q}\left[\nu\frac{\partial k}{\partial x_q}\right]}_{III}$$

$$\text{Eddy diffusion process }{}^kP_{IV}\;:\;\underbrace{\frac{\partial}{\partial x_q}\left[\sigma^*\nu_T\frac{\partial k}{\partial x_q}\right]}_{IV}$$

Like the transport equation of k, the intention of the modellers of the k-ω model is to add a transport equation (a PDE) for ω, too. However, one realizes that, unlike k, to begin with, there is no exact equation of ω available with us. Thus, a partial-differential equation modeling the evolution of ω needs to be entirely postulated.

For such a postulation, the modeller borrows some guidelines from the modelled equation of k (8.77), itself.

Like the modelled equation of k, (8.22), it is proposed to have the presence of four essential processes in the evolution equation of ω. These four processes are: the production rate of ω (${}^{\omega}P_I$), the dissipation-rate of ω (${}^{\omega}P_{II}$), the molecular diffusion rate of ω (${}^{\omega}P_{III}$), and eddy diffusion rate of ω (${}^{\omega}P_{IV}$). Thus, symbolically, the postulated transport equation of ω is expressed as:

$$\frac{\partial \omega}{\partial t} + \langle V_q \rangle \frac{\partial \omega}{\partial x_q} = {}^{\omega}P_I - {}^{\omega}P_{II} + {}^{\omega}P_{III} + {}^{\omega}P_{IV} \tag{8.25}$$

Further, we assume (a modeling assumption) that the respective functional form of the processes of the ω equation is the same as the corresponding processes of the k equation such that

$$\begin{aligned}
\text{Production process:} \quad {}^{\omega}P_I &= \left(\frac{\alpha\omega}{k}\right) {}^{k}P_I = \left(\frac{\alpha\omega}{k}\right) \frac{R_{ij}}{\rho} \frac{\partial \langle V_i \rangle}{\partial x_j} \\
\text{Dissipation process:} \quad {}^{\omega}P_{II} &= \left(\frac{\alpha' \omega}{k}\right) {}^{k}P_{II} = \left(\frac{\alpha' \omega}{k}\right) (\beta^* k \omega) \\
&= \beta \omega^2 \\
\text{Molecular diffusion process:} \quad {}^{\omega}P_{III} &= \frac{\partial}{\partial x_q} \left(\nu \frac{\partial \omega}{\partial x_q} \right) \\
\text{Eddy diffusion process:} \quad {}^{\omega}P_{IV} &= \frac{\partial}{\partial x_q} \left(\sigma \nu_T \frac{\partial \omega}{\partial x_q} \right)
\end{aligned} \tag{8.26}$$

where the factor ω/k has been introduced merely with the motivation to ensure the dimensional consistency of the production and the dissipation-rate processes of the ω equation. Further, new dimensionless constants (α, β and σ) have also been introduced to possibly have better control over the performance of the final closure model. Thus, the fully modelled transport equation of ω is:

$$\begin{aligned}
\frac{\partial \omega}{\partial t} + \langle V_q \rangle \frac{\partial \omega}{\partial x_q} &= \left(\frac{\alpha\omega}{k}\right) \frac{R_{ij}}{\rho} \frac{\partial \langle V_i \rangle}{\partial x_j} - \beta \omega^2 + \frac{\partial}{\partial x_q} \left(\nu \frac{\partial \omega}{\partial x_q} \right) \\
&+ \frac{\partial}{\partial x_q} \left(\sigma \nu_T \frac{\partial \omega}{\partial x_q} \right)
\end{aligned} \tag{8.27}$$

In summary, (5.2), (5.8), (8.5), (8.6), (8.72), (8.22) and (8.27) form a mathematically closed set of 18 scalar equations (a combination of algebraic equations and partial differential equations) employing the k-ω closure.

$$
\begin{aligned}
&\frac{\partial \langle V_i \rangle}{\partial x_i} = 0 \\
&\frac{\partial \langle V_i \rangle}{\partial t} + \langle V_k \rangle \frac{\partial \langle V_i \rangle}{\partial x_k} = -\frac{1}{\rho}\frac{\partial \langle p \rangle}{\partial x_i} + \nu \frac{\partial^2 \langle V_i \rangle}{\partial x_k \partial x_k} + \frac{1}{\rho}\frac{\partial R_{ki}}{\partial x_k} \\
&R_{ij} = R_{ij}^{\text{anisotropic}} - \frac{2}{3}\rho k \delta_{ij} \\
&R_{ij}^{\text{anisotropic}} = 2\mu_T \left\langle S_{ij} \right\rangle \\
&\mu_T = \rho \frac{k}{\omega} \\
&\frac{\partial k}{\partial t} + \left\langle V_q \right\rangle \frac{\partial k}{\partial x_q} = \frac{R_{ij}}{\rho}\frac{\partial \langle V_i \rangle}{\partial x_j} - \beta^* k \omega + \frac{\partial}{\partial x_q}\left[\left(\nu + \sigma^* \nu_T\right) \frac{\partial k}{\partial x_q} \right] \\
&\frac{\partial \omega}{\partial t} + \left\langle V_q \right\rangle \frac{\partial \omega}{\partial x_q} = \left(\frac{\alpha \omega}{k}\right) \frac{R_{ij}}{\rho}\frac{\partial \langle V_i \rangle}{\partial x_j} - \beta \omega^2 + \frac{\partial}{\partial x_q}\left[(\nu + \sigma \nu_T) \frac{\partial \omega}{\partial x_q} \right]
\end{aligned}
\tag{8.28}
$$

The 18 unknowns appearing in this equation set are: V_i (3 scalars), p (1 scalar), R_{ij} (6 scalars), $R_{ij}^{\text{anisotropic}}$ (5 scalars), μ_T (1 scalar), k (1 scalar) and ω (1 scalar). Further, there are five closure coefficients α, β^*, β, σ and σ^* that need to be prescribed by the modeler. Since, for closure, the k-ω model has added two additional partial differential equations (PDEs) (those of k and ω) over and above the four PDEs that are the RANS continuity and the momentum equations (5.2, and 5.8), this kind of closure is also called a *two-equation turbulence closure model.*

8.3.1 Selection of Closure Coefficients

In Sect. 8.2 we discussed that turbulence modellers tend to employ available experimental databases or direct numerical simulation databases of simple flow fields to select the best possible values of various closure coefficients. The specific choice of those simple flow fields depends on the modeller's preference. To select appropriate values of the closure coefficients, the k-ω model employs experimental/DNS databases of two simple flow fields: decaying turbulence and the log-layer of a flat plate boundary layer. The detailed procedure for selecting the closure coefficients is explained in the next subsections.

Decaying Turbulence

For the k-ω model, we first employ the database of decaying turbulence to select an appropriate value of the ratio β^*/β such that the model can reproduce some specific behavior as observed in a DNS/experimental database of such a flow field. In a decaying turbulence flow field (a statistically homogeneous flow field), spatial derivatives of all expected values vanish, leading to the following simplified forms of the modelled transport equations of k and ω (8.28).

$$\frac{dk}{dt} = -\beta^* k\omega \tag{8.29}$$

and

$$\frac{d\omega}{dt} = -\beta\omega^2 \tag{8.30}$$

where $\frac{\partial}{\partial t}$ has been replaced by $\frac{d}{dt}$, because neither k nor ω varies spatially.

The form of the ω Eq. (8.29) allows us to integrate it between two time instants t_o (a reference time in the past) and t (the current time instant) leading to the following explicit expression of ω in terms of t.

$$\omega(t) = \frac{\omega_o}{1 + \omega_o\beta(t - t_o)} \tag{8.31}$$

where $\omega_o = \omega(t = t_o)$.

Next, we use this explicit expression of $\omega(t)$ in (8.29) leading to the following form of the k equation.

$$\frac{dk}{dt} = -\beta^* k\omega = \frac{-\beta^* k\omega_o}{1 + \omega_o\beta(t - t_o)} \tag{8.32}$$

Integration of (8.32) between the same two-time instants t_o and t leads to the following expression.

$$k(t) = k_o\left(1 + \omega_o\beta(t - t_o)\right)^{-\beta^*/\beta} \tag{8.33}$$

where $k_o = k(t = t_o)$.

Having obtained explicit expressions for both $k(t)$ and $\omega(t)$, we now focus on a sufficiently late stage of the evolution of the flow field such that $t \gg t_o$. At such time instants, the expression in (8.33) is simplified.

$$\begin{aligned} k(t) &= k_o\left(1 + \omega_o\beta(t - t_o)\right)^{-\beta^*/\beta} \approx k_o\left(1 + \omega_o\beta t\right)^{-\beta^*/\beta} \\ &\approx k_o\left[(\omega_o\beta)^{-\beta^*/\beta}\right] t^{-\beta^*/\beta} \end{aligned} \tag{8.34}$$

Investigations employing an experimental database of decaying turbulence (grid turbulence) show that turbulence kinetic energy decays as

$$k \sim t^{-n} \tag{8.35}$$

where n is found to lie between 1.0 and 1.5 [6]. Comparing the expressions in (8.34) and (8.35), the modelers concluded

$$\frac{\beta^*}{\beta} = n. \tag{8.36}$$

Finally, the modelers selected 1.2 as the value of n leading to

$$\frac{\beta^*}{\beta} = 1.2. \tag{8.37}$$

Log Layer of a Flat Plate Boundary Layer

Proceeding further, the k-ω model is taken to another simple flow field: the flat plate boundary layer. We have discussed earlier in Chap. 6 that at high Re_1, the mean continuity and the mean momentum equation of the RANS equation (8.28) set can be simplified to the following form (already included in (6.77)).

$$\frac{\partial \langle V_1 \rangle}{\partial x_1} + \frac{\partial \langle V_2 \rangle}{\partial x_2} = 0,$$
$$\langle V_1 \rangle \frac{\partial \langle V_1 \rangle}{\partial x_1} + \langle V_2 \rangle \frac{\partial \langle V_1 \rangle}{\partial x_2} = \frac{-1}{\rho}\frac{\partial \langle p \rangle}{\partial x_1} + \nu \frac{\partial^2 \langle V_1 \rangle}{\partial x_2 \partial x_2} - \frac{1}{\rho}\frac{\partial \left\langle \rho V_1^{'} V_2^{'} \right\rangle}{\partial x_2} \tag{8.38}$$

The $\langle V_1 \rangle$ equation in (8.38) can alternatively be expressed in terms of $\sigma_{12}^{\text{mean}}$ as

$$\langle V_1 \rangle \frac{\partial \langle V_1 \rangle}{\partial x_1} + \langle V_2 \rangle \frac{\partial \langle V_1 \rangle}{\partial x_2} = \frac{1}{\rho}\frac{\partial (\sigma_{12}^{\text{mean}})}{\partial x_2} \tag{8.39}$$

where $\underline{\sigma}^{\text{mean}}$ is the total stress tensor in the mean flow field inside a boundary layer (6.80). The stress $\sigma_{12}^{\text{mean}}$ is composed of two parts: the viscous stress and the Reynolds stress

$$\sigma_{12}^{\text{mean}} = \mu \frac{\partial \langle V_1 \rangle}{\partial x_2} + \left\langle -\rho V_1^{'} V_2^{'} \right\rangle = \mu \frac{\partial \langle V_1 \rangle}{\partial x_2} + R_{12} \tag{8.40}$$

Using the closure suggested by the k-ω model (8.28), R_{12} is expressed as

$$R_{12} = -\left\langle \rho V_1^{'} V_2^{'} \right\rangle = 2\mu_T \langle S_{12} \rangle = 2\mu_T \left(\frac{\partial \langle V_1 \rangle}{\partial x_2} + \frac{\partial \langle V_2 \rangle}{\partial x_1} \right) \frac{1}{2}. \tag{8.41}$$

The order-of-magnitude analysis at $Re_1 \gg 1$ shows (6.31) that

$$\frac{\partial \langle V_1 \rangle}{\partial x_2} \gg \frac{\partial \langle V_2 \rangle}{\partial x_1} \tag{8.42}$$

Thus, (8.41) simplifies to

$$R_{12} = -\left\langle \rho V_1^{'} V_2^{'} \right\rangle \approx \mu_T \frac{\partial \langle V_1 \rangle}{\partial x_2}. \tag{8.43}$$

Thus, inside a boundary layer, the total stress, $\sigma_{12}^{\text{mean}}$ (8.40) is expressed as

$$\sigma_{12}^{\text{mean}} = \mu \frac{\partial \langle V_1 \rangle}{\partial x_2} + \left\langle -\rho V_1^{'} V_2^{'} \right\rangle = (\mu + \mu_T) \frac{\partial \langle V_1 \rangle}{\partial x_2} \tag{8.44}$$

Next, we refer to Fig. 6.5 (DNS database of a flat plate boundary layer) where we observe that inside the log layer

$$\sigma_{12}^{\text{mean}} \approx R_{12} \tag{8.45}$$

Further, as shown in Fig. 6.6, inside the log layer,

$$R_{12} \approx \tau_w(\text{ constant at a fixed axial station}) \tag{8.46}$$

Thus, based on the expression of $\sigma_{12}^{\text{mean}}$ in (8.44) and using these two specific observations (8.45 and 8.46), two conclusions are drawn for the flow inside the log layer:

$$\mu_T \gg \mu \tag{8.47}$$

and

$$\frac{\partial}{\partial x_2}\left(\mu_T \frac{\partial \langle V_1 \rangle}{\partial x_2}\right) = 0. \tag{8.48}$$

The DNS-based observations listed in Eqs. (8.45) and (8.46) also imply that the entire RHS of (8.39) vanishes. This, in turn, leads us to the conclusion that the LHS of (8.39), which represents the advection process of the quantity $\langle V_1 \rangle$, too, vanishes inside the log layer.

$$\langle V_1 \rangle \frac{\partial \langle V_1 \rangle}{\partial x_1} + \langle V_2 \rangle \frac{\partial \langle V_1 \rangle}{\partial x_2} \approx 0 \tag{8.49}$$

We now turn our attention to the modelled equations of k and ω (8.28). Since the flow inside a boundary layer is (i) statistically stationary, (ii) statistical homogeneous in the $\hat{e}_3$ direction and (iii) the gradient of an expected value in the $\hat{e}_2$ direction is much larger than that in the $\hat{e}_1$ direction, these equations simplify to the following forms.

$$\langle V_1 \rangle \frac{\partial k}{\partial x_1} + \langle V_2 \rangle \frac{\partial k}{\partial x_2} = \frac{R_{12}}{\rho} \frac{\partial \langle V_1 \rangle}{\partial x_2} - \beta^* k\omega + \frac{\partial}{\partial x_2}\left[\left(\nu + \sigma^* \nu_T\right) \frac{\partial k}{\partial x_2}\right] \tag{8.50}$$

$$\langle V_1 \rangle \frac{\partial \omega}{\partial x_1} + \langle V_2 \rangle \frac{\partial \omega}{\partial x_2} = \frac{\alpha\omega R_{12}}{\rho k} \frac{\partial \langle V_1 \rangle}{\partial x_2} - \beta\omega^2 + \frac{\partial}{\partial x_2}\left[(\nu + \sigma \nu_T) \frac{\partial \omega}{\partial x_2}\right] \tag{8.51}$$

where using the conclusions of the order-of-magnitude analysis (6.31 and 6.34) performed earlier in Chap. 6, the mean velocity gradient has been approximated to the following form.

$$\underline{\nabla} \langle \underline{V} \rangle \approx \frac{\partial \langle V_1 \rangle}{\partial x_2} \hat{e}_2 \hat{e}_1. \tag{8.52}$$

Next, we confine our discussion only to the *log layer*. Inside the log layer, we have $\mu_T \gg \mu$ (8.47). The symbols σ^* and σ appearing in the eddy diffusion processes of the two quantities are closure coefficients. Even though they have not been determined yet, we do expect them to have their orders of magnitude close to unity. With this anticipation, it is still plausible to neglect the viscous diffusion process in comparison to the eddy diffusion process in the two Eqs. (8.50 and 8.51). Thus,

$$\left[(\nu + \sigma^* \nu_T)\frac{\partial k}{\partial x_2}\right] \approx \left[\sigma^* \nu_T \frac{\partial k}{\partial x_2}\right]$$
$$\left[(\nu + \sigma \nu_T)\frac{\partial \omega}{\partial x_2}\right] \approx \left[\sigma \nu_T \frac{\partial \omega}{\partial x_2}\right]. \tag{8.53}$$

Accordingly, (8.50) and (8.51) are simplified further to the following forms.

$$\langle V_1 \rangle \frac{\partial k}{\partial x_1} + \langle V_2 \rangle \frac{\partial k}{\partial x_2} = \frac{R_{12}}{\rho} \frac{\partial \langle V_1 \rangle}{\partial x_2} - \beta^* k \omega + \frac{\partial}{\partial x_2}\left[\sigma^* \nu_T \frac{\partial k}{\partial x_2}\right] \tag{8.54}$$

$$\langle V_1 \rangle \frac{\partial \omega}{\partial x_1} + \langle V_2 \rangle \frac{\partial \omega}{\partial x_2} = \frac{\alpha \omega R_{12}}{\rho k} \frac{\partial \langle V_1 \rangle}{\partial x_2} - \beta \omega^2 + \frac{\partial}{\partial x_2}\left[\sigma \nu_T \frac{\partial \omega}{\partial x_2}\right] \tag{8.55}$$

Earlier, DNS-based evidence led us to the conclusion that the advection process of the $\langle V_1 \rangle$ must vanish inside the log layer. Even though at this point, there is no such direct evidence available for drawing similar conclusions about the respective advection processes of (8.54) and (8.55), it is assumed [10] that in the log layer, the advection processes of (8.54) and (8.55), too are negligible. This leads to the following simplified forms of (8.54) and (8.55).

$$0 = \frac{R_{12}}{\rho} \frac{\partial \langle V_1 \rangle}{\partial x_2} - \beta^* k \omega + \frac{\partial}{\partial x_2}\left[\sigma^* \nu_T \frac{\partial k}{\partial x_2}\right] \tag{8.56}$$

$$0 = \frac{\alpha \omega R_{12}}{\rho k} \frac{\partial \langle V_1 \rangle}{\partial x_2} - \beta \omega^2 + \frac{\partial}{\partial x_2}\left[\sigma \nu_T \frac{\partial \omega}{\partial x_2}\right] \tag{8.57}$$

Thus, the log-layer analyses of the $\langle V_1 \rangle$, k and ω transport equations adopted by the k-ω model has led us to a system of three Eqs. (8.48), (8.56) and (8.57) which are summarized below.

$$
\begin{aligned}
\frac{\partial}{\partial x_2}\left(\frac{k}{\omega}\frac{\partial \langle V_1\rangle}{\partial x_2}\right) &= 0\\
\frac{k}{\omega}\left[\frac{\partial \langle V_1\rangle}{\partial x_2}\right]^2 - \beta^* k\omega + \frac{\partial}{\partial x_2}\left[\sigma^*\frac{k}{\omega}\frac{\partial k}{\partial x_2}\right] &= 0\\
\alpha\left[\frac{\partial \langle V_1\rangle}{\partial x_2}\right]^2 - \beta\omega^2 + \frac{\partial}{\partial x_2}\left[\sigma\frac{k}{\omega}\frac{\partial \omega}{\partial x_2}\right] &= 0
\end{aligned}
\tag{8.58}
$$

where R_{12} has been expressed in terms of μ_T, and the quantity μ_T, itself has been expressed in terms of k and ω (8.28).

We observe (8.58) is a set of three equations in as many flow variables ($\langle V_1\rangle$, k and ω). It can be verified that the following algebraic functions of these flow variables do satisfy (8.58)

$$\langle V_1\rangle = \sqrt{\frac{\tau_w}{\rho}}\frac{1}{\kappa}ln\left\{\frac{x_2}{\nu}\sqrt{\frac{\tau_w}{\rho}}\right\} \tag{8.59}$$

$$k = \frac{u_\tau^2}{\sqrt{\beta^*}} \tag{8.60}$$

$$\omega = \frac{u_\tau}{\sqrt{\beta^*}\kappa x_2} \tag{8.61}$$

with an additional algebraic constraint on the closure coefficients (8.62),

$$\alpha = \frac{\beta}{\beta^*} - \frac{\sigma\kappa^2}{\sqrt{\beta^*}} \tag{8.62}$$

Algebraically, the constant κ appearing in the solution set can be arbitrarily chosen. However, to ensure the consistency of the algebraic solution of $\langle V_1\rangle$ (8.59) with the experimentally observed *law of the wall*, we must set it to be the same constant ($\kappa = 0.41$) that appears in the experimental curve fit shown in Fig. 6.4.

To leverage this solution set further for selecting closure coefficients of the k-ω model, we appeal an additional observation which is based on experimental data. An optimal curve fit in the log layer shows that [12]

$$\frac{\tau_w}{\rho k} \approx 0.3 \tag{8.63}$$

which, with a minor rearrangement, is expressed as

$$k \approx \frac{10}{3}u_\tau^2 \tag{8.64}$$

Comparing this experimentally-observed curve fit included in (8.64) with (8.60), we select

$$\beta^* = \frac{9}{100}. \tag{8.65}$$

Subsequently, using (8.37), we select

$$\beta = \frac{5}{6} \text{ and } \beta^* = \frac{3}{40}. \tag{8.66}$$

It seems the values of the two closure coefficients σ^* and σ, which are used to model the eddy diffusion process in the k and the ω equations, respectively, have been selected by the modellers [11] following a trial-based approach rather than any plausible physics-based arguments.

$$\sigma^* = \sigma = \frac{1}{2}. \tag{8.67}$$

The modeller justified these choices by comparing the performance of the k-ω with some experimentally observed behavior (no further details not available in [11], though. However, once these values have been selected, the constraint obtained from the log-layer analysis allows us to readily select a value of α.

$$\alpha = \frac{\beta}{\beta^*} - \frac{\sigma\kappa^2}{\sqrt{\beta^*}} = \frac{5}{9} \tag{8.68}$$

This step completes the selection of all closure coefficients, and the k-ω model is now summarized as:

$$\begin{aligned}
&\frac{\partial \langle V_i \rangle}{\partial x_i} = 0\\
&\frac{\partial \langle V_i \rangle}{\partial t} + \langle V_k \rangle \frac{\partial \langle V_i \rangle}{\partial x_k} = -\frac{1}{\rho}\frac{\partial \langle p \rangle}{\partial x_i} + \nu \frac{\partial^2 \langle V_i \rangle}{\partial x_k \partial x_k} + \frac{1}{\rho}\frac{\partial R_{ki}}{\partial x_k}\\
&R_{ij} = R_{ij}^{\text{anisotropic}} - \frac{2}{3}\rho k \delta_{ij}\\
&R_{ij}^{\text{anisotropic}} = 2\mu_T \left\langle S_{ij} \right\rangle\\
&\mu_T = \rho \frac{k}{\omega}\\
&\frac{\partial k}{\partial t} + \left\langle V_q \right\rangle \frac{\partial k}{\partial x_q} = \frac{R_{ij}}{\rho}\frac{\partial \langle V_i \rangle}{\partial x_j} - \beta^* k\omega + \frac{\partial}{\partial x_q}\left[\left(\nu + \sigma^* \nu_T\right)\frac{\partial k}{\partial x_q}\right]\\
&\frac{\partial \omega}{\partial t} + \left\langle V_q \right\rangle \frac{\partial \omega}{\partial x_q} = \left(\frac{\alpha\omega}{k}\right)\frac{R_{ij}}{\rho}\frac{\partial \langle V_i \rangle}{\partial x_j} - \beta\omega^2 + \frac{\partial}{\partial x_q}\left[(\nu + \sigma \nu_T)\frac{\partial \omega}{\partial x_q}\right]\\
&\beta^* = \frac{9}{100}, \beta = \frac{3}{40}, \sigma^* = \frac{1}{2}, \sigma = \frac{1}{2}, \alpha = \frac{5}{9}.
\end{aligned} \tag{8.69}$$

For more information about various computational aspects and performance benchmarks related to the k-ω model, the reader is referred to [11].

8.4 The k-ϵ Model: A Case Study on Turbulence Modeling

The present form of the k-ϵ model is based on the work of Jones and Launder [13] and Launder and Sharma [14]. To model μ_T (8.7), like the k-ω model, first the k-ϵ model chooses to estimate $V_{\text{fluctation}}$ using the local instantaneous value of k in the flow field,

$$V_{\text{fluctation}} \sim k^{1/2}. \tag{8.70}$$

Further, the model introduces the dissipation rate (ϵ) to estimate $l_{\text{fluctuation}}$ by using (7.96).

$$\epsilon \sim \frac{u_o^3}{l_o} \sim \frac{k^{3/2}}{l_{\text{fluctuation}}} \Rightarrow l_{\text{flctuation}} \sim \frac{k^{3/2}}{\epsilon} \tag{8.71}$$

Using (8.71) in (8.7), we arrive at the following modeled form of μ_T

$$\mu_T = C_\mu \rho \frac{k^2}{\epsilon} \tag{8.72}$$

where C_μ is a dimensionless constant introduced by the modellers. Thus, the two unclosed unknowns in our hands are k and ϵ. In general, we expect them to change with space and time. The k-ϵ model proposes to add (to the existing set of four partial differential equations, 5.10), two additional partial differential equations describing the evolution of k and ϵ in the flow field.

We have earlier derived the transport equation of k (5.56). The unclosed terms are highlighted in (8.12) as A, B and C, and these need to be modelled. Since the modellers are willing to add a transport equation for ϵ, anyway, they leave the dissipation rate ϵ as it is in Eq. (8.12).

To model Term B in (8.12), like the k-ω model, the gradient diffusion hypothesis is adopted (8.21), and Term B is modelled in the following form.

$$\underbrace{\frac{\partial}{\partial x_q}\left\langle V_q' \frac{V_i' V_i'}{2} \right\rangle}_{B} = -\frac{\partial}{\partial x_q}\left(\frac{\gamma k^2}{\epsilon}\frac{\partial k}{\partial x_q}\right) \tag{8.73}$$

where γ is a dimensionless constant.

Term C is the next unclosed term in (8.12). The main modeling challenge posed by this term is the involvement of pressure fluctuations in it. Again like the k-ω model, it is decided to drop this term altogether, because it poses a challenge due to the presence of pressure fluctuations therein. In summary, the modelled form of the transport equation of k for the k-ϵ model is as follows.

$$
\begin{aligned}
\frac{\partial k}{\partial t}+\left\langle V_q\right\rangle \frac{\partial k}{\partial x_q} &= \frac{R_{ij}}{\rho}\frac{\partial\left\langle V_i\right\rangle}{\partial x_j}+\nu\frac{\partial^2 k}{\partial x_q \partial x_q}-\epsilon+\frac{\partial}{\partial x_q}\left(\frac{C_\mu k^2}{\sigma_k \epsilon}\frac{\partial k}{\partial x_q}\right) \\
&= \underbrace{\frac{R_{ij}}{\rho}\frac{\partial\left\langle V_i\right\rangle}{\partial x_j}}_{I}-\underbrace{\epsilon}_{II}+\underbrace{\frac{\partial}{\partial x_q}\left[\nu\frac{\partial k}{\partial x_q}\right]}_{III}+\underbrace{\frac{\partial}{\partial x_q}\left[\frac{\nu_T}{\sigma_k}\frac{\partial k}{\partial x_q}\right]}_{IV}
\end{aligned}
\tag{8.74}
$$

where ν_T is the kinematic eddy viscosity,

$$
\nu_T = C_\mu \frac{k^2}{\epsilon} \tag{8.75}
$$

The dimensionless constant γ appearing in (8.73) has been combined with C_μ to have a new closure coefficient σ_k appearing in the eddy diffusion process, and this remains to be determined. Equation (8.22) is symbolically expressed as

$$
\frac{\partial k}{\partial t}+\left\langle V_q\right\rangle \frac{\partial k}{\partial x_q} = {}^kP_I - {}^kP_{II} + {}^kP_{III} + {}^kP_{IV} \tag{8.76}
$$

where the four individual processes on the RHS of this equation are identified as

$$
\text{Production process }\ {}^kP_I = \underbrace{\frac{R_{ij}}{\rho}\frac{\partial\left\langle V_i\right\rangle}{\partial x_j}}_{I}
$$

$$
\text{Dissipation process }\ {}^kP_{II} : \underbrace{\epsilon}_{II}
$$

$$
\text{Molecular diffusion process }\ {}^kP_{III} : \underbrace{\frac{\partial}{\partial x_q}\left[\nu\frac{\partial k}{\partial x_q}\right]}_{III}
$$

$$
\text{Eddy diffusion process }\ {}^kP_{IV} : \underbrace{\frac{\partial}{\partial x_q}\left[\frac{\nu_T}{\sigma_k}\frac{\partial k}{\partial x_q}\right]}_{IV}
$$

Like the transport equation of k, the intention of the modelers of the k-ϵ was to add a transport equation for ϵ, too. The exact equation of ϵ can be derived. However, the resulting equation (not included here) has multiple unclosed terms, which require modeling anyway. Thus, a partial-differential equation modeling the evolution of ϵ is postulated in its entirety, and for such a postulation, we borrow some guidelines from the modelled equation of k (8.76).

Like the modelled equation of k, (8.76), we propose to have the presence of four essential processes in the evolution equation of ϵ. These four processes are: the production rate of ϵ (${}^\epsilon P_I$), the dissipation-rate of ϵ (${}^\epsilon P_{II}$), the molecular diffusion rate of ϵ (${}^{\epsilon}P_{III}$), and the eddy diffusion rate of ϵ (${}^{\epsilon}P_{IV}$). Thus, symbolically, the transport equation of ϵ is expressed as:

$$\frac{\partial \epsilon}{\partial t} + \langle V_q \rangle \frac{\partial \epsilon}{\partial x_q} = {}^{\epsilon}P_I - {}^{\epsilon}P_{II} + {}^{\epsilon}P_{III} + {}^{\epsilon}P_{IV} \tag{8.77}$$

Further, we assume (a modeling assumption) that the respective functional form of the processes of the ϵ equation are the same as the corresponding processes of the k Eq. (8.76) such that

$$\begin{aligned}
\text{Production process: } {}^{\epsilon}P_I &= \left(\frac{C_{\epsilon 1}\epsilon}{k}\right) {}^{k}P_I \\
&= \left(\frac{C_{\epsilon 1}\epsilon}{k}\right) \frac{R_{ij}}{\rho} \frac{\partial \langle V_i \rangle}{\partial x_j} \\
\text{Dissipation process: } {}^{\epsilon}P_{II} &= \left(\frac{C_{\epsilon 2}\epsilon}{k}\right) {}^{k}P_{II} = \frac{C_{\epsilon 2}\epsilon^2}{k} \\
\text{Molecular diffusion process: } {}^{\epsilon}P_{III} &= \frac{\partial}{\partial x_q}\left(\nu \frac{\partial \epsilon}{\partial x_q}\right) \\
\text{Eddy diffusion process: } {}^{\epsilon}P_{IV} &= \frac{\partial}{\partial x_q}\left(\frac{\nu_T}{\sigma_\epsilon} \frac{\partial \epsilon}{\partial x_q}\right)
\end{aligned} \tag{8.78}$$

where the factor ϵ/k has been introduced merely with the motivation to ensure the dimensional consistency of the production and the dissipation-rate processes of the ϵ equation. Further, new dimensionless constants ($C_{\epsilon 1}$, $C_{\epsilon 2}$ and σ_ϵ) have also been introduced to possibly have better control over the performance of the final closure model. Thus, the final form of the fully modelled transport equation of ϵ is:

$$\begin{aligned}
\frac{\partial \epsilon}{\partial t} + \langle V_q \rangle \frac{\partial \epsilon}{\partial x_q} = &\left(\frac{C_{\epsilon 1}\epsilon}{k}\right) \frac{R_{ij}}{\rho} \frac{\partial \langle V_i \rangle}{\partial x_j} - C_{\epsilon 2}\frac{\epsilon^2}{k} + \frac{\partial}{\partial x_q}\left(\nu \frac{\partial \epsilon}{\partial x_q}\right) \\
&+ \frac{\partial}{\partial x_q}\left(\frac{\nu_T}{\sigma_\epsilon} \frac{\partial \epsilon}{\partial x_q}\right)
\end{aligned} \tag{8.79}$$

In summary, (5.2), (5.8), (8.5), (8.6), (8.72), (8.74) and (8.79) form a mathematically closed set of 18 scalar equations (a combination of algebraic equations and partial differential equations) employing the k-ϵ closure.

$$\frac{\partial \langle V_i \rangle}{\partial x_i} = 0$$

$$\frac{\partial \langle V_i \rangle}{\partial t} + \langle V_k \rangle \frac{\partial \langle V_i \rangle}{\partial x_k} = -\frac{1}{\rho}\frac{\partial \langle p \rangle}{\partial x_i} + \nu \frac{\partial^2 \langle V_i \rangle}{\partial x_k \partial x_k} + \frac{1}{\rho}\frac{\partial R_{ki}}{\partial x_k}$$

$$R_{ij} = R_{ij}^{\text{anisotropic}} - \frac{2}{3}\rho k \delta_{ij}$$

$$
\begin{aligned}
R_{ij}^{\text{anisotropic}} &= 2\mu_T \left\langle S_{ij} \right\rangle \\
\mu_T &= C_\mu \frac{\rho k^2}{\epsilon} \\
\frac{\partial k}{\partial t} + \left\langle V_q \right\rangle \frac{\partial k}{\partial x_q} &= \frac{R_{ij}}{\rho} \frac{\partial \left\langle V_i \right\rangle}{\partial x_j} - \epsilon + \frac{\partial}{\partial x_q} \left[\left(\nu + \frac{\nu_T}{\sigma_k} \right) \frac{\partial k}{\partial x_q} \right] \\
\frac{\partial \epsilon}{\partial t} + \left\langle V_q \right\rangle \frac{\partial \epsilon}{\partial x_q} &= \left(\frac{C_{\epsilon 1} \epsilon}{k} \right) \frac{R_{ij}}{\rho} \frac{\partial \left\langle V_i \right\rangle}{\partial x_j} - \frac{C_{\epsilon 2} \epsilon^2}{k} + \frac{\partial}{\partial x_q} \left[\left(\nu + \frac{\nu_T}{\sigma_\epsilon} \right) \frac{\partial \epsilon}{\partial x_q} \right]
\end{aligned}
\tag{8.80}
$$

The 18 unknowns of in this equation set are: V_i (3 scalars), p (1 scalar), R_{ij} (6 scalars), $R_{ij}^{\text{anisotropic}}$ (5 scalars), μ_T (1 scalar), k (1 scalar) and ϵ (1 scalar). Further, there are five closure coefficients C_μ, σ_k, $C_{\epsilon 1}$, $C_{\epsilon 2}$ and σ_ϵ that need to be selected by the modeler. Since, for closure, the k-ϵ model has added two additional partial differential equations (PDEs) (those of k and ϵ) over and above the four PDEs that are the RANS continuity and the momentum equations, like the k-ω model, this is another example of the two-equation turbulence closure models.

8.4.1 Selection of Closure Coefficients

To select appropriate values of the closure coefficients, the k-ϵ model employs experimental/DNS databases of three simple flow fields: decaying turbulence, the log-layer of a flat plate boundary layer and homogeneous shear flows. The detailed procedure for selecting the closure coefficients is explained in the next subsections.

Decaying Turbulence

For the k-ϵ model, the modellers first employ the database of decaying turbulence to select an appropriate value of $C_{\epsilon 2}$, such that the model can reproduce some specific behavior as observed in experimental databases of such a flow field. In this flow field (a statistically homogeneous flow field), spatial derivatives of all expected values vanish, leading to the following simplified forms of modelled transport equation of k and ϵ (8.80).

$$\frac{dk}{dt} = -\epsilon \tag{8.81}$$

and

$$\frac{d\epsilon}{dt} = -C_{\epsilon 2} \frac{\epsilon^2}{k} \tag{8.82}$$

where $\frac{\partial}{\partial t}$ has been replaced by $\frac{d}{dt}$ because neither k nor ϵ varies spatially (decaying turbulence has a statistically homogeneous flow field). It can be verified that (8.81) and (8.82) are simultaneously satisfied by the following functions of k and ϵ in terms of t (time)

$$k(t) = \frac{k_o}{\alpha_o^{-n}} t^{-n} \text{ and } \epsilon(t) = \frac{\epsilon_o}{\alpha_o^{-n-1}} t^{-n-1}, \tag{8.83}$$

subject to the condition

$$C_{\epsilon 2} = \frac{n+1}{n} \tag{8.84}$$

The symbols k_o and ϵ_o are the values of k and ϵ at a reference time. The symbol α_o is related to n (a non-dimensional number) in the following manner.

$$n = \frac{\alpha_o \epsilon_o}{k_o}. \tag{8.85}$$

It is observed in experiments [6] on decaying turbulence, that indeed k decays as $k \sim t^{-n}$ (in accordance with (8.83)) where n is observed to lie within the range of 1.0 and 1.5 in experiments [6]. The k-ϵ modeler selected $n = 1.08$. Thus, (8.84) leads to

$$C_{\epsilon 2} = 1.92 \tag{8.86}$$

Log Layer of a Flat Plate Boundary Layer

Proceeding further with the selection of the values of the closure coefficients, the k-ϵ model is taken to the log layer of the flat plate boundary layer. Following exactly the same arguments as presented in (8.38)–(8.46), in context of the k-ω model, we are again led to the same conclusions that in the log-layer,

$$\mu_T \gg \mu \tag{8.87}$$

and

$$\frac{\partial}{\partial x_2}\left(\mu_T \frac{\partial \langle V_1 \rangle}{\partial x_2}\right) = 0, \tag{8.88}$$

which for the k-ϵ model (8.88) means

$$\frac{\partial}{\partial x_2}\left(\mu_T \frac{\partial \langle V_1 \rangle}{\partial x_2}\right) = \frac{\partial}{\partial x_2}\left(\frac{\rho C_\mu k^2}{\epsilon} \frac{\partial \langle V_1 \rangle}{\partial x_2}\right) = 0, \tag{8.89}$$

where the expression of μ_T has been used from (8.72).

The DNS-based observations listed in Eqs. (8.45) and (8.46) also imply that the entire RHS of (8.39) vanishes. This, in turn, leads us to the conclusion that the LHS of (8.39), which represents the advection process of the quantity $\langle V_1 \rangle$, too, vanishes inside the log layer.

$$\langle V_1 \rangle \frac{\partial \langle V_1 \rangle}{\partial x_1} + \langle V_2 \rangle \frac{\partial \langle V_1 \rangle}{\partial x_2} \approx 0 \tag{8.90}$$

We now turn our attention to the modelled equations of k and ϵ (8.80). Since the flow inside a boundary layer is (i) statistically stationary, (ii) statistical homogeneous in the $\hat{e}_3$ direction and (iii) the gradient of any expected value in the $\hat{e}_2$ direction is much larger than that in the $\hat{e}_1$ direction, these equations simplify to the following forms.

$$\langle V_1\rangle \frac{\partial k}{\partial x_1} + \langle V_2\rangle \frac{\partial k}{\partial x_2} = \frac{R_{12}}{\rho}\frac{\partial \langle V_1\rangle}{\partial x_2} - \epsilon + \frac{\partial}{\partial x_2}\left[\left(\nu + \frac{\nu_T}{\sigma_k}\right)\frac{\partial k}{\partial x_2}\right] \tag{8.91}$$

$$\langle V_1\rangle \frac{\partial \epsilon}{\partial x_1} + \langle V_2\rangle \frac{\partial \epsilon}{\partial x_2} = \frac{C_{\epsilon 1}\epsilon R_{12}}{\rho k}\frac{\partial \langle V_1\rangle}{\partial x_2} - \frac{C_{\epsilon 2}\epsilon^2}{k} + \frac{\partial}{\partial x_2}\left[\left(\nu + \frac{\nu_T}{\sigma_\epsilon}\right)\frac{\partial \epsilon}{\partial x_2}\right] \tag{8.92}$$

where using the conclusions of the order-of-magnitude analysis (6.31 and 6.34) performed earlier in Chap. 6, the mean velocity gradient has been approximated to the following form.

$$\underline{\nabla}\langle \underline{V}\rangle \approx \frac{\partial \langle V_1\rangle}{\partial x_2}\hat{e}_2\hat{e}_1. \tag{8.93}$$

Next, we confine our discussion to the *log layer*. Inside the log layer, we have $\mu_T \gg \mu$ (8.47). Even though the closure coefficients σ_k and σ_ϵ have not been determined yet, it is anticipated to have their orders of magnitude close to unity. With this anticipation, it is still plausible to neglect the viscous diffusion process in comparison to the eddy diffusion process in the two equations. Thus,

$$\begin{aligned}\left[\left(\nu + \frac{\nu_T}{\sigma_k}\right)\frac{\partial k}{\partial x_2}\right] &\approx \left[\frac{\nu_T}{\sigma_k}\frac{\partial k}{\partial x_2}\right]\\ \left[\left(\nu + \frac{\nu_T}{\sigma_\epsilon}\right)\frac{\partial \epsilon}{\partial x_2}\right] &\approx \left[\frac{\nu_T}{\sigma_\epsilon}\frac{\partial \epsilon}{\partial x_2}\right].\end{aligned} \tag{8.94}$$

Accordingly, (8.94) further simplifies (8.91) and (8.92) to the following forms.

$$\langle V_1\rangle \frac{\partial k}{\partial x_1} + \langle V_2\rangle \frac{\partial k}{\partial x_2} = \frac{R_{12}}{\rho}\frac{\partial \langle V_1\rangle}{\partial x_2} - \epsilon + \frac{\partial}{\partial x_2}\left[\frac{\nu_T}{\sigma_k}\frac{\partial k}{\partial x_2}\right] \tag{8.95}$$

$$\langle V_1\rangle \frac{\partial \epsilon}{\partial x_1} + \langle V_2\rangle \frac{\partial \epsilon}{\partial x_2} = \frac{C_{\epsilon 1}\epsilon R_{12}}{\rho k}\frac{\partial \langle V_1\rangle}{\partial x_2} - \frac{C_{\epsilon 2}\epsilon^2}{k} + \frac{\partial}{\partial x_2}\left[\frac{\nu_T}{\sigma_\epsilon}\frac{\partial \epsilon}{\partial x_2}\right] \tag{8.96}$$

Earlier, DNS-based evidence led us to the conclusion that the advection process of the $\langle V_1 \rangle$ must vanish inside the log layer. Even though at this point, there is no such direct evidence available for drawing such conclusions about (8.95) and (8.96), it is assumed [11] that in the log layer, the advection processes are negligible for k and ϵ, as well. This leads to the following simplified forms of the two Eqs. (8.95) and (8.96).

$$0 = \frac{R_{12}}{\rho} \frac{\partial \langle V_1 \rangle}{\partial x_2} - \epsilon + \frac{\partial}{\partial x_2} \left[\frac{\nu_T}{\sigma_k} \frac{\partial k}{\partial x_2} \right] \tag{8.97}$$

$$0 = \frac{C_{\epsilon 1} \epsilon R_{12}}{\rho k} \frac{\partial \langle V_1 \rangle}{\partial x_2} - \frac{C_{\epsilon 2} \epsilon^2}{k} + \frac{\partial}{\partial x_2} \left[\frac{\nu_T}{\sigma_\epsilon} \frac{\partial \epsilon}{\partial x_2} \right] \tag{8.98}$$

Thus, the log-layer analyses of the $\langle V_1 \rangle$, k and ϵ transport equations proposed by the k-ϵ model has led us to a system of three Eqs. (8.89), (8.97) and (8.98) which are summarized below.

$$\begin{aligned} \frac{\partial}{\partial x_2} \left(\frac{C_\mu k^2}{\epsilon} \frac{\partial \langle V_1 \rangle}{\partial x_2} \right) &= 0 \\ \frac{C_\mu k^2}{\epsilon} \left[\frac{\partial \langle V_1 \rangle}{\partial x_2} \right]^2 - \epsilon + \frac{\partial}{\partial x_2} \left[\frac{C_\mu k^2}{\epsilon \sigma_k} \frac{\partial k}{\partial x_2} \right] &= 0 \\ C_{\epsilon 1} C_\mu k \left[\frac{\partial \langle V_1 \rangle}{\partial x_2} \right]^2 - \frac{C_{\epsilon 2} \epsilon^2}{k} + \frac{\partial}{\partial x_2} \left[\frac{C_\mu k^2}{\epsilon \sigma_\epsilon} \frac{\partial \epsilon}{\partial x_2} \right] &= 0 \end{aligned} \tag{8.99}$$

where R_{12} has been expressed in terms of μ_T, and the quantity μ_T, itself has been expressed in terms of k and ϵ.

We observe (8.99) is a set of three equations in as many flow variables ($\langle V_1 \rangle$, k and ϵ). It can be verified that the following algebraic functions of these flow variables do satisfy (8.99)

$$\langle V_1 \rangle = \sqrt{\frac{\tau_w}{\rho}} \frac{1}{\kappa} ln \left\{ \frac{x_2}{\nu} \sqrt{\frac{\tau_w}{\rho}} \right\} \tag{8.100}$$

$$k = \frac{u_\tau^2}{\sqrt{C_\mu}} \tag{8.101}$$

$$\epsilon = \frac{u_\tau^3}{\kappa x_2} \tag{8.102}$$

with the following additional algebraic constraint on the closure coefficients,

$$\kappa^2 = \sigma_\epsilon \left(C_{\epsilon 2} - C_{\epsilon 1} \right) \sqrt{C_\mu} \tag{8.103}$$

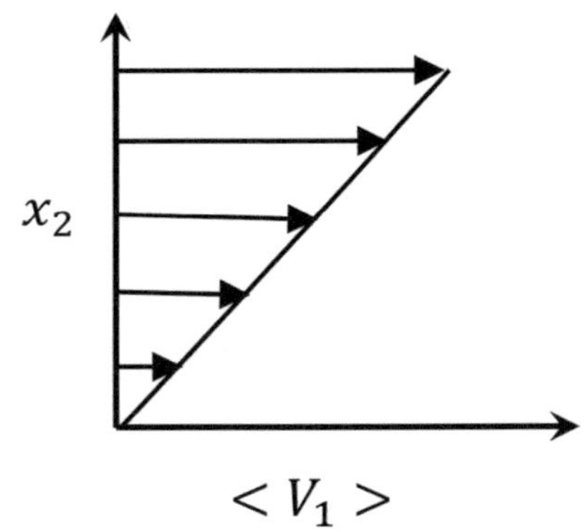

Fig. 8.3 Profile of $\langle V_1 \rangle$ in a homogeneous shear flow

Algebraically, the constant κ appearing in this solution set can be arbitrarily chosen. However, to ensure consistency of the algebraic solution of $\langle V_1 \rangle$ (8.100) with the experimentally observed *law of the wall* (6.90), we set that to be the same constant that appears in the experimental curve fit of the *law of the wall* ($\kappa = 0.41$).

To leverage this solution set further for selecting the closure coefficients of the k-ϵ model, we appeal to another observation made on some experimental data of a flat plate turbulent boundary layer. An optimal curve fit in the log layer shows that [12]

$$\frac{\tau_w}{\rho k} \approx 0.3 \tag{8.104}$$

which, with a minor rearrangement, is expressed as

$$k \approx \frac{10}{3} u_\tau^2 \tag{8.105}$$

Comparing this experimentally-observed curve fit with (8.101), we select

$$C_\mu = \frac{9}{100}. \tag{8.106}$$

Homogeneous Shear Flow

The modellers of the k-ϵ model employ some known behavior of another kind of simple flow field—*homogeneous shear flow*—to select the closure coefficient σ_ϵ. Before we provide the details of this selection procedure, we briefly describe a homogeneous shear flow.

A *homogeneous shear flow* is a turbulent flow field with the mean velocity gradient being of the following form.

$$\underline{\nabla} \langle \underline{V} \rangle = \mathcal{S} \hat{e}_2 \hat{e}_1 \tag{8.107}$$

where $\mathcal{S}$ is a constant. In Fig. 8.3 we present a schematic of the mean velocity profile in a homogeneous shear flow. The slope of the velocity profile equals $\mathcal{S}$ (8.107). Further, a homogeneous shear flow has *homogeneous turbulence*, which we described earlier in Chap. 5 (5.69). A homogeneous shear flow can be set up in a direct numerical simulation [15] or in a wind tunnel experiment [16], and various statistics of interest can be computed or measured. The utility of the homogeneous shear flow is that it allows us to investigate the exact interaction between the production and the dissipation processes of the turbulence kinetic energy without the interfering effects of any inhomogeneous transport process (5.70). In a homogeneous shear flow, the exact transport equation of k (5.56) simplifies (without making any further assumptions) to

$$\frac{dk}{dt} = {}^k\mathcal{P} - \epsilon \tag{8.108}$$

where

$${}^k\mathcal{P} = \frac{R_{12}}{\rho}\frac{\partial\langle V_1\rangle}{\partial x_2} = \frac{R_{12}\mathcal{S}}{\rho} \tag{8.109}$$

and

$$\epsilon = \nu\left\langle \frac{\partial V_i'}{\partial x_j}\frac{\partial V_i'}{\partial x_j}\right\rangle \tag{8.110}$$

Several DNS databases of homogeneous shear flow shows that at large normalized time instants, $\mathcal{S}t$, both k/ϵ and ${}^k\mathcal{P}/\epsilon$ tend to attain time-dependent asymptotic values in these simulations.

$$\frac{d}{dt}\left(\frac{k}{\epsilon}\right) \approx 0 \text{ at } \mathcal{S}t \gg 0 \tag{8.111}$$

and

$$\frac{d}{dt}\left(\frac{{}^k\mathcal{P}}{\epsilon}\right) \approx 0 \text{ and } \left(\frac{{}^k\mathcal{P}}{\epsilon}\right) \to 1.8 \text{ at } \mathcal{S}t \gg 0 \tag{8.112}$$

We now turn our attention to the modelled k and ϵ Eq. (8.80). In a homogeneous shear flow, these equations are simplified (all inhomogeneous processes must vanish)

$$\begin{aligned} \frac{dk}{dt} &= {}^k\mathcal{P} - \epsilon \\ \frac{d\epsilon}{dt} &= \left(\frac{C_{\epsilon 1}\epsilon}{k}\right){}^k\mathcal{P} - \frac{C_{\epsilon 2}\epsilon^2}{k} \end{aligned} \tag{8.113}$$

With these equations, we find the time derivative of k/ϵ.

$$\begin{aligned} \frac{d}{dt}\left(\frac{k}{\epsilon}\right) &= \frac{1}{\epsilon}\frac{dk}{dt} - \frac{k}{\epsilon^2}\frac{d\epsilon}{dt} \\ &= \frac{{}^k\mathcal{P}}{\epsilon} - 1 - \frac{C_{\epsilon 1}{}^k\mathcal{P}}{\epsilon} + C_{\epsilon 2} \end{aligned} \tag{8.114}$$

This relationship is valid for the k-ϵ model at all times in a homogeneous shear flow. However, our interest is to focus on the asymptotic behavior when the LHS of (8.114) vanishes, leading to the following equation for $C_{\epsilon 1}$

$$C_{\epsilon 1} = (C_{\epsilon 2} - 1)\frac{\epsilon}{k\mathcal{P}} + 1 \tag{8.115}$$

Even though in direct numerical simulations, the asymptotic value is observed to be $\frac{k\mathcal{P}}{\epsilon} = 1.8$, the k-ϵ modeller employed $\frac{k\mathcal{P}}{\epsilon} = 2.1$ to arrive at

$$C_{\epsilon 1} = 1.44. \tag{8.116}$$

Having selected the values of C_μ (8.106), $C_{\epsilon 2}$ (8.86) and $C_{\epsilon 1}$ (8.116), we go back to the algebraic constraint that emerged from the log-layer analysis (8.103). This leads to arrive at a value of σ_ϵ

$$\sigma_\epsilon = \frac{\kappa^2}{\sqrt{C_\mu}(C_{\epsilon 2} - C_{\epsilon 1})} = 1.3 \tag{8.117}$$

where the modeller used $\kappa = 0.43$.

Lastly, the value of the closure coefficients σ_k, which has been used to model the eddy diffusion process in the k equation, has been chosen as (without any physics-based justification)

$$\sigma_k = 1. \tag{8.118}$$

This step (8.118) completes the selection of all closure coefficients. Thus, the k-ϵ model, along with all closure coefficients, is now summarized as:

$$\begin{aligned}
&\frac{\partial \langle V_i \rangle}{\partial x_i} = 0 \\
&\frac{\partial \langle V_i \rangle}{\partial t} + \langle V_k \rangle \frac{\partial \langle V_i \rangle}{\partial x_k} = -\frac{1}{\rho}\frac{\partial \langle p \rangle}{\partial x_i} + \nu \frac{\partial^2 \langle V_i \rangle}{\partial x_k \partial x_k} + \frac{1}{\rho}\frac{\partial R_{ki}}{\partial x_k} \\
&R_{ij} = R_{ij}^{\text{anisotropic}} - \frac{2}{3}\rho k \delta_{ij} \\
&R_{ij}^{\text{anisotropic}} = 2\mu_T \langle S_{ij} \rangle \\
&\mu_T = C_\mu \frac{\rho k^2}{\epsilon} \\
&\frac{\partial k}{\partial t} + \langle V_q \rangle \frac{\partial k}{\partial x_q} = \frac{R_{ij}}{\rho}\frac{\partial \langle V_i \rangle}{\partial x_j} - \epsilon + \frac{\partial}{\partial x_q}\left[\left(\nu + \frac{\nu_T}{\sigma_k}\right)\frac{\partial k}{\partial x_q}\right] \\
&\frac{\partial \epsilon}{\partial t} + \langle V_q \rangle \frac{\partial \epsilon}{\partial x_q} = \left(\frac{C_{\epsilon 1}\epsilon}{k}\right)\frac{R_{ij}}{\rho}\frac{\partial \langle V_i \rangle}{\partial x_j} - \frac{C_{\epsilon 2}\epsilon^2}{k} + \frac{\partial}{\partial x_q}\left[\left(\nu + \frac{\nu_T}{\sigma_\epsilon}\right)\frac{\partial \epsilon}{\partial x_q}\right] \\
&C_\mu = \frac{9}{100}, \sigma_k = 1, C_{\epsilon 1} = 1.44, C_{\epsilon 2} = 1.92, \sigma_\epsilon = 1.3
\end{aligned} \tag{8.119}$$

One of the major drawbacks of the k-ϵ model is that the modelled transport equation of ϵ does not correctly show the near-wall asymptotic behavior (discussed earlier in Chap. 6) in the viscous sublayer. Thus, this form of the transport equation of ϵ cannot be integrated all the way to a solid wall inside a boundary layer. Over the years, many modified forms of the k-ϵ model have been proposed which can address this problem of the model presented in (8.119). To differentiate the model presented in (8.119) from those modified versions, the former is commonly referred to as the *standard* k-ϵ model. For further information about various computational aspects and the performance benchmarks related to the k-ϵ model, the reader is referred to [11].

Scale-Resolving Simulations of Turbulent Flows

9

In Chap. 5 we introduced the RANS equation set to attain the ability to seek solution of the set of primary variables of engineering interest: $\langle \underline{V_i} \rangle$ and $\langle \underline{p} \rangle$. If we could attain this ability, this would have paved the way for finding the expected values of the velocity and the pressure fields without relying on any type of averaging. However, the RANS equation set (5.10) turned out to be mathematically unclosed due to the appearance of the Reynolds stress tensor in the mean momentum equation. Many turbulence closure models have been proposed to achieve this mathematical closure. However, these closure models are approximate by their very nature, and they invariably add uncertainty to the solution of the mean velocity and the mean pressure fields that emerge using these closure models.

On the other hand, direct numerical simulations, which solve the exact governing equations of the velocity and the pressure fields (3.9), do not require any closure model, and thus they are free of any such modeling uncertainty. If applied to a statistically stationary or a flow field which has at least one direction of statistical homogeneity, the solution of the instantaneous flow field from a direct numerical simulation can be appropriately averaged to obtain the expected values of the velocity and the pressure fields. However, the main challenge posed by a direct numerical simulation is the requirement of having extremely fine resolution in temporal and spatial discretization. This makes it prohibitive to perform DNS of flows at high Reynolds numbers.

The Kolmogorov hypotheses, which we discussed in Chap. 7, suggest that the small-scale motion in a turbulent flow field tends to be more isotropic and universal. In contrast, the large-scale motion tends to be more dependent on the boundary conditions of the flow field. Further, we also realize that it is the requirement of resolving the smallest scales of motion (the Kolmogorov length and time scales), rather than the larger scales of motion, which actually intensifies the computational

S. S. Sinha, *Fundamentals of Turbulence and Its Modeling*,
https://doi.org/10.1007/978-3-031-94016-3_9

demand of performing a direct numerical simulation. These two realizations lead us to the idea of conceptually decomposing the velocity and the pressure fields based on a filter function.

9.1 The Filtered Description of a Turbulent Flow Field

In the filtered description, the instantaneous flow variables are subjected to a filter function which tends to retain the contributions of only those scales which are larger than a cutoff length-scale (say L_{cutoff}) and discards the contribution of the scales smaller than L_{cutoff} to the flow variable of interest. We symbolically denote such a filtering operation by $\langle . \rangle_f$. Accordingly, the instantaneous velocity and pressure variables are decomposed as

$$V_i = \langle V_i \rangle_f + V_i^{''} \text{ and } p = \langle p \rangle_f + p^{''} \tag{9.1}$$

The symbols $V_i^{''}$ and $p^{''}$ are called the residues of V_i and p subject to the filtering process. These residues, by definition, when added back to their filtered counterparts, recover the corresponding instantaneous variables.

We anticipate that seeking numerical solutions of $\langle V_i \rangle_f$ and $\langle p \rangle_f$ would pose less stringent requirements of spatial and temporal resolution to the CFD methodology compared to resolving the instantaneous variables V_i and p themselves in a direct numerical simulation. Equation (9.1) forms the basis of what we call the *scale-resolving simulations*. Strictly speaking, a direct numerical simulation is the most accurate scale-resolving simulation. However, in contemporary literature, the phrase *scale-resolving simulations* is typically used to refer to those numerical simulations which attempt to solve for the filtered velocity and pressure fields ($\langle V_i \rangle_f$ and $\langle p \rangle_f$).

The decomposition (9.1) is apparently similar to the decomposition introduced in Chap. 5 wherein the instantaneous velocity and pressure variables were decomposed in terms of the means and the fluctuations.

$$V_i = \langle V_i \rangle + V_i^{'} \text{ and } p = \langle p \rangle + p^{'} \tag{9.2}$$

However, unlike the mean operator, $\langle \rangle$, with which

$$\langle \langle \phi \rangle \rangle = 0 \text{ and } \left\langle \phi^{'} \right\rangle = 0 \tag{9.3}$$

for the filtering operator, $\langle \rangle_f$, in general,

$$\left\langle \langle \phi \rangle_f \right\rangle_f \neq \langle \phi \rangle_f \text{ and } \left\langle \phi_i^{''} \right\rangle_f \neq 0 \tag{9.4}$$

where ϕ is an instantaneous variable of a turbulent flow field.

Further, unlike $\langle \phi \rangle$, which is a deterministic quantity, $\langle \phi \rangle_f$ is still a random variable. Thus, to extract any meaningful statistics of the random instantaneous flow variable (ϕ), an appropriate averaging process of $\langle \phi \rangle_f$ would still be required.

9.2 Governing Equations of a Filtered Flow Field

The governing equations of the filtered flow field are the governing equations of our primary variables of interest in a scale-resolving simulation: $\langle V_i \rangle_f$ and $\langle p \rangle_f$. These equations are derived from the instantaneous Navier-stocks equation set (3.10 and 3.11). At this point, we assume the filter function $\langle . \rangle_f$ to have the following properties,

$$\left\langle \frac{\partial \phi_1}{\partial t} \right\rangle_f = \frac{\partial \langle \phi_1 \rangle_f}{\partial t},$$
$$\left\langle \frac{\partial \phi_1}{\partial x_i} \right\rangle_f = \frac{\partial \langle \phi_1 \rangle_f}{\partial x_i},$$
$$\text{and} \quad \langle \phi_1 + \phi_2 \rangle_f = \langle \phi_1 \rangle_f + \langle \phi_2 \rangle_f \tag{9.5}$$

where ϕ_1 and ϕ_2 are two instantaneous flow variables of interest.

We subject the instantaneous continuity equation (3.10) to the filtering operator.

$$\left\langle \frac{\partial V_i}{\partial x_i} \right\rangle_f = \langle 0 \rangle_f \text{ or,} \tag{9.6}$$

subsequently, using the properties listed in (9.5) leads us to the following form of the *filtered continuity equation*

$$\frac{\partial \langle V_i \rangle_f}{\partial x_i} = 0 \tag{9.7}$$

Next, we subject the instantaneous momentum equation (3.11) to the filtering operator

$$\left\langle \frac{\rho \partial V_i}{\partial t} + \rho V_k \frac{\partial V_i}{\partial x_k} \right\rangle_f = \left\langle -\frac{\partial p}{\partial x_i} + \mu \frac{\partial^2 V_i}{\partial x_k \partial x_k} \right\rangle_f \tag{9.8}$$

Again, employing the properties listed in (9.5), the filtered momentum equation is expressed as

$$\frac{\partial \langle \rho V_i \rangle_f}{\partial t} + \left\langle \rho V_k \frac{\partial V_i}{\partial x_k} \right\rangle_f = -\frac{\partial \langle p \rangle_f}{\partial x_i} + \mu \frac{\partial^2 \langle V_i \rangle_f}{\partial x_k \partial x_k} \tag{9.9}$$

Since ρ is a constant, it commutes across both the filtering and the derivative operators.

$$\rho \frac{\partial \langle V_i \rangle_f}{\partial t} + \rho \left\langle V_k \frac{\partial V_i}{\partial x_k} \right\rangle_f = -\frac{\partial \langle p \rangle_f}{\partial x_i} + \mu \frac{\partial^2 \langle V_i \rangle_f}{\partial x_k \partial x_k} \tag{9.10}$$

The second term on the LHS Eq. (9.10) is the filtered value of the advection term. Using the instantaneous continuity equation (3.10), this term is expressed as

$$\rho\left\langle V_k \frac{\partial V_i}{\partial x_k}\right\rangle_f = \rho\left\langle \frac{\partial\left(V_i V_k\right)}{\partial x_k} - V_i \frac{\partial V_k}{\partial x_k}\right\rangle_f = \rho\left\langle \frac{\partial\left(V_i V_k\right)}{\partial x_k}\right\rangle_f$$
$$= \rho\frac{\partial\left\langle V_i V_k\right\rangle_f}{\partial x_k} \tag{9.11}$$

Further, using the decomposition scheme (9.2), we recast the filtered value of the product of velocity components as

$$\begin{aligned}\left\langle V_i V_k\right\rangle_f &= \left\langle\left(\left\langle V_i\right\rangle_f + V_i''\right)\left(\left\langle V_k\right\rangle_f + V_k''\right)\right\rangle_f \\ &= \left\langle\left\langle V_i\right\rangle_f\left\langle V_k\right\rangle_f + V_i''\left\langle V_k\right\rangle_f + V_k''\left\langle V_i\right\rangle_f + V_i'' V_k''\right\rangle_f \\ &= \left\langle\left\langle V_i\right\rangle_f\left\langle V_k\right\rangle_f\right\rangle_f + \left\langle V_i''\left\langle V_k\right\rangle_f\right\rangle + \left\langle V_k''\left\langle V_i\right\rangle_f\right\rangle_f \\ &+ \left\langle\left\langle V_i'' V_k''\right\rangle_f\right\rangle_f\end{aligned} \tag{9.12}$$

All four terms on the RHS of (9.12) are, in general, non-zero. Further, all these quantities are deemed as *secondary unknowns*.

It is conventional to club together all the terms on the RHS of (9.12) and express (9.12) as the product of the filtered velocity components $\langle V_i\rangle_f$ and $\langle V_k\rangle_f$ and additionally one single secondary unknown quantity $\underline{\tau}_{ik}^{T}$:

$$\left\langle V_i V_k\right\rangle_f = \left\langle V_i\right\rangle_f\left\langle V_k\right\rangle_f + \tau_{ik}^{T} \tag{9.13}$$

where τ_{ik}^{T} represents the (ik)th component of the second order tensor $\underline{\tau}^{T}$, which is called the *generalized central moment* tensor of the velocity field [17]. Alternatively, it is also referred to as the (ik)th component of the *turbulent stress tensor*.

Substituting (9.13) in (9.10) leads to the following form of equation:

$$\frac{\partial\left\langle V_i\right\rangle_f}{\partial t} + \frac{\partial\left(\left\langle V_k\right\rangle_f\left\langle V_i\right\rangle_f\right)}{\partial x_k} = -\frac{1}{\rho}\frac{\partial\langle p\rangle_f}{\partial x_i} + \nu\frac{\partial^2\left\langle V_i\right\rangle_f}{\partial x_k \partial x_k} - \frac{\partial \tau_{ik}^{T}}{\partial x_k} \tag{9.14}$$

Using the filtered continuity equation (9.7), the advection term in (9.14) can be further simplified and (9.14) is expressed as

$$\frac{\partial\left\langle V_i\right\rangle_f}{\partial t} + \left\langle V_k\right\rangle_f\frac{\partial\left\langle V_i\right\rangle_f}{\partial x_k} = -\frac{1}{\rho}\frac{\partial\langle p\rangle_f}{\partial x_i} + \nu\frac{\partial^2\left\langle V_i\right\rangle_f}{\partial x_k \partial x_k} - \frac{\partial \tau_{ik}^{T}}{\partial x_k} \tag{9.15}$$

Equation (9.15) is called the *filtered momentum equation*. The set of equations comprised of (9.7) and (9.15) is called the *filtered Navier-Stokes equation* (summarized below).

$$\frac{\partial \langle V_i \rangle_f}{\partial x_i} = 0$$

$$\frac{\partial \langle V_i \rangle_f}{\partial t} + \langle V_k \rangle_f \frac{\partial \langle V_i \rangle_f}{\partial x_k} = -\frac{1}{\rho}\frac{\partial \langle p \rangle_f}{\partial x_i} + \nu \frac{\partial^2 \langle V_i \rangle_f}{\partial x_k \partial x_k} - \frac{\partial \tau_{ik}^T}{\partial x_k} \tag{9.16}$$

The *primary unknowns* of the filtered Navier-Stokes equation set are the filtered velocity components and the filtered pressure variable: $\langle V_i \rangle_f$ and $\langle p \rangle_f$. The tensor $\underline{\tau}^T$ is deemed as a "new" quantity or *secondary* unknown tensor. This is a symmetric tensor and thus represents, in general, six secondary scalar unknowns. Like the instantaneous momentum equations (3.11), the mean momentum equations (9.15) represent three non-linear PDEs. The nonlinearity arises because of the advection term (second term on the LHS of 9.15).

Like the RANS equation set (5.10), the filtered Navier-Stokes equation (9.16) is mathematically unclosed, and closure modeling is required for the $\underline{\tau}^T$ tensor. However, if L_{cutoff} is chosen to be small and close to the dissipative range of scales of motion, the required models are expected to be simpler and more widely applicable. This anticipation is based on the Kolmogorov hypotheses, which suggest that the statistics pertaining to the smallest scales of motion tend to be more isotropic and universal.

The so-called *large eddy simulation or LES* methodology of turbulence computations sets the L_{cutoff} to be somewhat larger than the dissipative length-scales and then uses simple algebraic closure models for the τ_{ij}^T components. Since the L_{cutoff} is still larger than the smallest scale of motion (the Kolmogorov length-scale), an LES simulation can be performed on a grid coarser than what is required for a direct numerical simulation of the same flow field. This results into a considerable reduction in the requirement of computational resources. For more details about the LES methodology, the reader is referred to [6].

In recent years, some other scale-resolving methods have also been developed, which allow the user to choose L_{cutoff} anywhere between the range

$$L_{EI} > L_{\text{cutoff}} > L_{DI} \tag{9.17}$$

depending on the spatial resolution of the computational grid the user can afford to have. Such methods are, in general, called the *bridging methods* of turbulence computations. Even though bridging methods offer the freedom to the user to choose L_{cutoff} commensurate to the available computational resource, the simple algebraic closure models (like the ones used for LES) may not work effectively, and more advanced models involving additional partial differential equations are required for improved predictions of flow statistics. For more details on bridging methods, the reader is referred to [18].

Bibliography

1. S. S. Sinha. *Fundamentals of Fluid Mechanics*. Ane Books Pvt. Ltd, New Delhi, 2024.
2. S. Sengupta, P. C. Dumir, and S. V. Veeravalli. *Engineering Mechanics*. University Press (India) Private Limited, Hyderabad, India, 2020.
3. P. Schlatter and R. Örlu. "Assessment of direct numerical simulation data of turbulent boundary layers". In: *Journal of Fluid Mechanics* 659 (2010).
4. John D Anderson. "Ludwig Prandtl s boundary layer". In: *Physics today* 58.12 (2005).
5. A. E. Honein and P. Moin. "Higher entropy conservation and numerical stability of compressible turbulence simulations". In: *Journal of Computational Physics* 201 (2004).
6. S. B. Pope. *Turbulent Flows*. Cambridge University Press, 2000.
7. N. Parashar, S. S. Sinha, and B. Srinivasan. "Lagrangian investigations of velocity gradients in compressible turbulence: lifetime of flow-field topologies". In: *Journal of Fluid Mechanics* 872 (2019).
8. F. Waleffe. "The nature of triad interactions in homogenous turbulence". In: *Physics of Fluids* 4 (1992).
9. W.G. Vincenti and C. H. Kruger. *Introduction to Physical Gas Dynamics*. Krieger Pub Co, 1975.
10. David C Wilcox et al. *Turbulence modeling for CFD*. Vol. 2. DCW industries La Canada, CA, 1998.
11. D. C. Wilcox. "Reassessment of the scale-determining equation for advanced turbulence models". In: *AIAA Journal* 26.11 (1988).
12. A. A. R. Townsend. *The Structure of Turbulent Shear Flow*. Cambridge University Press, 1976, pp. –.
13. W. P. Jones and B. E. Launder. "The prediction of laminarization with a two-equation model of turbulence". In: *International Journal of Heat and Mass Tranfer* 15 (1972).
14. B.E. Launder and Sharma B. I. "Application of the energy-dissipation model of turbulence to the calculation of flow near a spinning disc". In: *Lett. Heat and Mass Transfer* 1 (1974).
15. R. S. Rogallo. *Numerical experiments in homogenous turbulence*. NASA Technical Memorandum 81315, 1981.
16. S. Tovoularis and S. Corrsin. "Experiments in nearly homogenous turbulent shear flow with a uniform mean temperature gradient. Part 1". In: *Journal of Fluid Mechanics* 104 (1981).
17. Massimo Germano. "Turbulence: the filtering approach". In: *Journal of Fluid Mechanics* 238 (1992), pp. 325–336.
18. S.S. Girimaji. "Partially-Averaged Navier-Stokes Model for Turbulence: A Reynolds-Averaged Navier-Stokes to Direct Numerical Simulation Bridging Method". In: *Journal of Applied Mechanics* 73 (2006).

S. S. Sinha, *Fundamentals of Turbulence and Its Modeling*,
https://doi.org/10.1007/978-3-031-94016-3

Index

A

advection operator, 14
anisotropic part, 5, 136
antisymmetric tensor, 5
area averaging, 54
averaged angular velocity vector, 26

B

Bernoulli equation, 95
boundary layer thickness, 78
bridging methods, 169
buffer layer, 100

C

central moment, 50
chaos, 37
characteristic length scale of an eddy, 29
characteristic time scale of an eddy, 29
characteristic value, 82
closure coefficients, 140
coefficient of diffusion, 142
continuum description, 19
covariance, 51
critical Reynolds number, 38
cumulative distribution function (CDF), 41
curl of a vector, 14

D

decaying turbulence, 105, 156
direct numerical simulation, 37
dissipation process, 69
dissipation range, 129
divergence of a tensor, 13
dot product of two tensors, 9
double dot product of two tensors, 12
dummy index, 7
dyads, 4
dynamic eddy viscosity, 138

E

ϵ-δ identity, 13
eddy, 28
eddy turnover time, 120
eddy viscosity closure, 136
Einstein's summation rule, 7
energy cascade, 121
energy-containing range, 128
energy spectrum function, 118
ensemble averaging, 53
Eulerian description, 22
event, 39
expected value, 45

F

filtered continuity equation, 168
filtered momentum equation, 168
filter function, 166
fluctuating kinetic energy per unit mass, 73
fluctuation, 48
fluid element, 19
fluid particle, 19
Fourier amplitudes, 107

S. S. Sinha, *Fundamentals of Turbulence and Its Modeling*,
https://doi.org/10.1007/978-3-031-94016-3

Fourier description, 107
Fourier modes, 107
Fourier space, 118
free index, 8
free vortex, 27
friction velocity, 97
spectral space, 118

G
Gaussian random variable, 42
generalized central moment, 168
gradient diffusion hypothesis, 142
gradient of a tensor, 13
grid turbulence, 107

H
homogeneous shear flow, 160
homogeneous turbulence, 160
homogenous turbulence, 74

I
index notation, 9
inertial reference frame, 36
inertial subrange, 129
inhomogeneous processes, 74
instantaneous flow field, 37
isotropic part, 5

J
joint CDF, 43
joint PDF, 44

K
coefficient of kinematic eddy viscosity, 144
k-ω model, 141
Kolmogorov length-scale, 131
Kolmogorov's first similarity hypothesis, 130
Kolmogorov's hypotheses, 127
Kolmogorov's hypothesis of local isotropy, 133
Kolmogorov's second similarity hypothesis, 131
Kolmogorov time scale, 131
Kolmogorov velocity scale, 131
Kronecker delta, 11, 116

L
Lagrangian description, 21
Laplacian operator, 14
large eddy simulation, 169
law of the wall, 100, 157
line averaging, 54
log layer, 100, 157

M
material derivative, 23
mean, 45
mean continuity equation, 57
mean material derivative, 69
mean momentum equation, 58
mean transport equation, 69
molecular diffusion process, 69, 142
moments of a random variable, 49

N
Navier Stokes equation set, 36
near-wall asymptotic behavior, 101
non-uniform velocity field, 23
normal strain-rate components, 27

O
outer layer, 100

P
Parseval's theorem, 116
peculiar velocity, 136
permutation symbol, 12
Prandtl boundary layer equation set, 93
pressure-strain correlation process, 69
probability density function (PDF), 41
probability of an event, 40
probability theory, 39
production process, 69

R
random field, 39
random process, 39
random variables, 35
raw moment, 49
realization, 40
Reynolds number, 34, 88, 106
Reynolds stress tensor, 59
rotation-rate tensor, 27

S
sample space, 40
scale-Resolving simulations, 165
shear strain-rate components, 27
spectral density function of ϵ, 119
spectral density function of k, 117
standard deviation, 50
statistically homogeneous, 52
statistically homogeneous flow, 105
statistically stationary, 52

steady velocity field, 23
Stokes constitutive equation, 136
strain-rate tensor, 27
symmetric tensor, 5

T

Taylor microscale, 106
tensor of order n, 1
time-averaging, 53
trace of a tensor, 5
transformation rule: first-order, 2
transformation rule of second-order tensors, 6
transitional flow, 38
triadic process, 127
turbulence closure problem, 59
turbulence kinetic energy, 69
turbulent boundary layer, 77
two-component velocity field, 23
two-dimensional velocity field, 23
two-equation turbulence closure models, 156

U

u^+, 97
uniformly distributed random variable, 43
unsteady velocity field, 23

V

variance, 50
viscous sublayer, 100
volume averaging, 55, 116
vortex, 27
vorticity vector, 26

W

wall shear stress, 96
wavenumber vector, 108

Y

y^+, 96

MIX
Papier aus verantwortungsvollen Quellen
Paper from responsible sources
FSC® C105338

If you have any concerns about our products,
you can contact us on
ProductSafety@springernature.com

In case Publisher is established outside the EU,
the EU authorized representative is:
Springer Nature Customer Service Center GmbH
Europaplatz 3, 69115 Heidelberg, Germany

Printed by Libri Plureos GmbH
in Hamburg, Germany